Air Forces of the World

Two CF-5As of No 434 Squadron, Canadian Air Force. / Canadian Forces

Saab SF37 Viggen reconnaissance aircraft of the Swedish Air Force. / Saab-Scania

Air Forces of the World

Barry C. Wheeler

Charles Scribner's Sons
NEW YORK

1 3 5 7 9 11 13 15 17 19 I/C 20 18 16 14 12 10 8 6 4 2

Printed in the United States of America

Library of Congress Catalog Card Number 79-65509
ISBN 0-684-16286-5

Abbreviations

STOL Short take-off and landing
AF Air Force
AB Air Base
SAR Search and Rescue
ASW Anti-Submarine Warfare
AOP Air Observation Post
ECM Electronic Countermeasures
ASM Air-to-Surface Missile
ELINT Electronic intelligence
OCU Operational Conversion Unit
PR Photographic Reconnaissance
FAC Forward Air Control
TFW Tactical Fighter Wing
AEW Airborne Early Warning

Introduction

The protection of a nation's land, its people and its resources has been of paramount importance for more than a thousand years. Today, that protection is more than just an army and its weapons. It comprises in many cases sophisticated missile defence systems, supersonic combat aircraft, nuclear submarines and large surface fleets. The investment in current national defence is measured in millions and billions of dollars and the content of this book endeavours to show where exactly some of those dollars are being spent.

Military aviation is big business and the economies of many western countries rest to a lesser or greater extent on exports connected with this section of the arms trade. Following the large-scale re-equipment that occurred after 1945 with redundant wartime aircraft, many air arms maintained a gradual replacement programme through the 1950s and early 1960s until the advent of the great arms boom of the 1970s. Heralded by the almost unlimited cash from Arab and Iranian oil which prompted the acquisition of modern air defence systems undreamed of a decade before, the race for arms became a battle between East and West, tangible only in the equipment arriving in the Third World and the Middle East and the wars that inevitably resulted. By 1977, arms dealing around the world was roughly split between the United States with 40%, Russia with 20% and the United Kingdom with approximately 8%. Politics play a major role in the procurement of modern equipment and a country's future allegiance is often clarified when either MiGs or F-5s suddenly arrive for that country's air force. In the larger East-West scenario, the smaller nations are frequently wooed into providing strategic bases from which the respective Power can keep an eye on the other.

In the last 10 years, a major *volte face* has taken place in a number of areas. Egypt, a country served by Russia since the days of President Nasser, is now firmly in the western camp with the USA, Britain and France updating her air force. In Africa, nations once loyal to the west have succumbed to 'wars of liberation' with Cuban mercenaries in their thousands, ably assisted by Russia, conducting military operations in Angola and Ethiopia. Central and South American air forces, traditionally recipients of United States aid, are gradually moving towards other sources of new equipment, Israel, France and Russia having gained significant orders for new aircraft over the past few years. Turbulent Asia, with communist fighting communist in Vietnam and Cambodia, is seen by many western arms manufacturers as a lucrative source for new aircraft orders: India has ordered Jaguars, Indonesia Hawks, while China has evinced more than a passing interest in the Harrier strike aircraft.

However, in all these situations the effectiveness of an air force rests not only with its equipment, but in the men who fly the aircraft and others who form its policies. Sound flying and top quality combat training have shown themselves to be the deciding factors in war, hence the large number of technical support contracts which accompany orders for new aircraft. These ensure sophisticated weapons are not wasted and nationals are trained to understand and operate them effectively. As far as equipment is concerned, worldwide inflation has forced many smaller nations to cut their defence budgets and invest in cheaper, lightweight combat aircraft, capable of multi-role missions. Machines such as the Northrop F-5 and the GD F-16 are finding more and more customers as older, larger jets reach obsolescence.

The entries in this book have been arranged alphabetically for ease of reference, each one divided into two main categories, Combat elements and Transport and Support types. The 'Strength' figure at the heading of each entry is the total number of aircraft in service which is generally more reliable than a combat strength which can be misleading as many training aircraft are capable of conducting light strike duties, giving them a quasi combat status. Where a country has naval and army aviation components which are not large enough for separate entries, they have been included at the end of the main air force text. To maintain readability, certain air force unit designations have been simplified into squadrons, groups and wings while other, more accepted ones have been retained, such as the German Geschwader.

This is essentially an up-to-date record of air force equipment, strengths, and operational bases, but because of the inevitable book publication lead-time, the information is correct up to February 1979. While the material is believed to be correct, many air arms remain sensitive about their military equipment and have refused to comment on their entry. To ensure greater accuracy in future editions, the author would welcome any updated information; this should be forwarded to the publishers.

Many people have helped in supplying details and photographs for this book and thanks are due to the public relations officers of the main aircraft manufacturers, the air forces that supplied data and pictures, various enthusiast bodies and organisations, and a number of individuals who took the time to forward items of interest to the author. In particular thanks go to Rick Alexander, Albert Anido, J. Merchant, R. Saunders, G. Russell and M. Stroud. The photographs came from a number of sources including the personal collections of Harry Holmes, Mike Hooks, Simon Thomson and Steve Richards.

February 1979 Barry C. Wheeler

Left· *Two of the world's most widely used helicopters, an Agusta-Bell 205 (top) and an AB206 of the Sultan of Oman's Air Force.*

Afghanistan

Title: Afghan Air Force (De Afghan Hanoi Quiran)
Headquarters: Kabul
Strength: 226+ aircraft

Combat
The Soviet Union provides limited military aid to Afghanistan under an agreement dating back to 1955. The air force is wholly Russian-equipped and currently operates a fighter element of three squadrons of MiG-21s with approximately 40 aircraft based at Pagram near Kabul. These aircraft were actively employed during the coup of mid-1977 which saw the establishment of a strong pro-Soviet government. A wing of four squadrons with 50 MiG-17s, assigned the day interceptor role, operates from Mazar-i-Sharif in the north of the country. Two further squadrons have 24 Sukhoi Su-7BM strike aircraft, and another unit has 12 MiG-19s supplied in 1965. Three units are equipped with nearly 45 Ilyushin Il-28 light bombers.

Transport and Support
Mainstay of the transport force is the Ilyushin Il-14 and about 25 have been supplied to the air force by the Soviet Union. In addition there are two Ilyushin Il-18Ds used for VIP and Government flights, and 10 Antonov An-2 biplane transports. A single helicopter squadron operates 18 Mil Mi-4s and a small number of the more modern Mi-8s. Under Soviet guidance, training is performed on two-seat MiG-15UTIs, MiG-21Us and Il-28Us; a batch of Czech L-39 Albatross advanced trainers is believed to be replacing the older types in service.

Albania

Title: Albanian People's Army Air Force
Headquarters: Tirana
Strength: 130+ aircraft

Combat
The communist state of Albania is a signatory of the Warsaw Pact but severed its relations with Russia in 1961 and left the Pact in September 1968. Since then the country has relied on Communist China for military and economic aid but indications during 1978 showed a marked drop in Chinese assistance to the country. The air force is almost totally Chinese-equipped and operates nearly 20 Chinese-built MiG-21s, known as Shenyang F-8s, two interceptor squadrons with 36 Shenyang F-6s (MiG-19s licence-built in China), two fighter-bomber squadrons with 24 Shenyang F-4s (MiG-17Fs) and two squadrons with 24 Shenyang F-2s (MiG-15s).

Transport and Support
A transport unit has three Ilyushin Il-14s and about 10 Antonov An-2s, plus nearly 30 Mil Mi-4s in a further two squadrons. Trainers include Yak-18s for basic tuition and some two-seat MiG-15UTIs for advanced flying.

Algeria

Title: Force Aérienne Algérienne (Al Quwwat Aljawwiya Aljaza'eriiya)
Headquarters: Algiers
Strength: 328 aircraft

Established in 1962 in the wake of the French withdrawal from the country, the FAA has received substantial aid from the Soviet Union and Egypt, although more recently equipment has been ordered from both France and the United States. Algeria has a defence agreement with neighbouring Libya and is providing support for the Polisario Front guerilla movement against Mauretania and Morocco. The Algerian air force has a wing of three interceptor squadrons flying 70 MiG-21Fs, three fighter-bomber squadrons with 60 MiG-17Fs and two squadrons with 20 Sukhoi Su-7BM strike aircraft. A second-line unit has 20 MiG-15s, while two bomber squadrons, of doubtful operational status, are equipped with 24

Below: *Algerian Air Force Fokker F27.* / Fokker-VFW

Ilyushin Il-28s. In 1971, the air force received 28 refurbished ex-Luftwaffe Magister twin-jet trainers from Aérospatiale and these have a dual strike/trainer role. By late 1978 Algeria was reported to have received up to 40 MiG-23/27 aircraft from Russia.

Transport and Support

Forming the backbone of the two transport squadrons in the FAA are eight Soviet-supplied Antonov An-12 heavy-lift freighters supported by about six Il-14s, four Il-18 VIP transports, and six Fokker F27s comprising five Mk 400s and one Mk 600. One Beech King Air is flown on navaid-calibration duties, while for short-range liasion and training duties there are three Super King Airs and three older Queen Airs. The helicopter force is made up of five Aérospatiale SA330 Puma assault machines bought in 1971, six Hughes 269 trainers, 40 Mil Mi-4s, 12 Mi-8s and four heavy-lift Mi-6s. To update the training force, six Beech T-34C-1 Turbo-Mentors have been ordered and will supplement the Magisters at the basic stage.

The Algerian Navy has established a small aviation arm with the recent purchase of two Fokker-VFW F28 Mk 3000Cs for transport and VIP work.

Angola

Title: Angolan Republic Air Force
Headquarters: Luanda
Strength: 80+ aircraft

Combat

This ex-Portuguese colony is governed by the Marxist MPLA following a civil war in 1976. With Soviet and Cuban assistance a national air arm was formed towards the end of the war under the title FAPA, the nucleus being provided by a batch of 12 MiG-21MF fighter-bombers. These aircraft were operated in the ground-attack role against the western-backed FNLA and Unita forces, flown by Cuban-trained Portuguese mercenaries and Cubans. In addition, eight MiG-17Fs were supplied, together with three MiG-15UTI two-seat trainers. Also absorbed into the air force are three ex-Portuguese AF Fiat G91R-4 light attack jets left behind at Luanda when Portugal yielded the country to the Angolans.

Reports emanating from South Africa in mid-1978 stated that three squadrons of Soviet MiG-23 and MiG-19 fighter-bombers had been stationed at bases in the far south of the country, at Serpa Pinto and Lubango, apparently to deter future South African Air Force attacks in the area.

Transport and Support

A variety of types are in use with FAPA for transport and liaison duties including three Douglas C-47s, five of six Antonov An-26s (the sixth being shot down in July 1977 by guerrilla forces) supplied by Russia, two Pilatus Turbo-Porters delivered in 1976 and about 20 Alouette IIIs. Other types believed to be in service include some Do27s, North American T-6s and some Portuguese OGMA-built Austers.

Argentina

Title: Fuerza Aerea Argentina
Headquarters: Buenos Aires
Strength: 370 aircraft

Combat

The second largest republic in South America, Argentina has been prevented from spending large amounts on her armed forces by rapid inflation but still fields a respectable air force. The FAA's flying elements are operated within the responsibility of Air Operations Command, within which are five air brigades. Each brigade, which roughly equates to an RAF wing, has up to three squadrons with a nominal establishment of three four-aircraft elements. The Second Air Brigade (II Brigada Aerea), headquartered at General Urquiza air base, Entre Rios, controls I Escuadron de Bombardeo (nine BAC Canberra B62s and two T64 trainers delivered in 1970-71) and II Escuadron de Exploration y Ataque (25 FMA IA35

Below: *BAC Canberra B62 of the Argentine Air Force.* / BAC

Huanqueros and about 12 of the 30 FMA IA58 Pucara attack aircraft ordered). Delivery of Pucaras to Reconquista air base — home of II Escuadron, one of two squadrons scheduled to get the type — has been well behind schedule; the FAA's requirement stands at about 100 aircraft. The remaining II Brigade squadron is I Escuadron Fotografico, equipped with 20 FMA IA35s.

Operating under the auspices of IV Brigada Aerea is I Escuadron at El Plumerillo air base, Mendoza, equipped with 25 McDonnell Douglas A-4P Skyhawk fighter-bombers, and II and III Escuadron flying locally-built Morane MS760 Paris two-seat armed jet trainers. Skyhawks also form the main equipment of IV and V Escuadron de Caza-Bombardeo of the V Brigada Aerea at General Pringles air base, San Luis. The A-4P Skyhawks are refurbished US Navy A-4B/Cs delivered in 1966 and 1970; some 45 are in use with V Brigada, making a total inventory of some 70 aircraft. To update these machines, Ferranti D126R Isis weapons aiming sights are being fitted following an initial order placed in 1976; further orders are likely.

The VII Brigada Aerea at Moron air base, Buenos Aires, operates the FAA's sole interceptor unit — I Escuadron de Caza, flying Dassault Mirages. A total of 12 IIIEAs and two IIIDAs was delivered in 1972 and a further seven IIIEAs were ordered recently to make up for attrition; plans for the acquisition of a further 80 Mirages have been shelved for economic reasons. In addition, the brigade has a helicopter attack squadron, I Escuadron de Exploration y Ataque, equipped with 14 Hughes 500Ms and six Bell UH-1H gunships.

Transport and Support

Prime element of the transport units is I Escuadron of I Brigada Aerea, based at El Palomar, Buenos Aires, operating three Lockheed C-130Es and four C-130H Hercules (a fifth H was destroyed in 1975) delivered between 1968 and 1975. At the same base is the Dept de Aviones Presidenciales, a special VIP unit equipped with one HS748 Srs 2, one Fokker-VFW F28 Mk 1000C, one Rockwell Sabreliner and a Boeing 707-320B. A further Escuadron operates 22 FMA IA50 Guarani IIs including 14 transport versions, two survey machines and a VIP aircraft. El Palomar is also HQ of the state military airline LADE (Lineas Aereas del Estado), which operates passenger and cargo services to remote areas of the country. The airline's fleet comprises five Fokker-VFW F28 Mk 1000Cs, five F27 Mk 400s, five F27 Mk 600s,three Douglas DC-6Bs and seven DHC Twin Otters.

For short-haul liaison work there are 14 Rockwell Shrike Commanders and an SAR unit — Escuadron de Busqueda y Salvamento — equipped with three Grumman HU-16B Albatross amphibians for coastal duties and six Aérospatiale Lamas for high-altitude rescue. Fixed-wing aeromedical flights are performed by two Swearingen Merlin IVAs delivered in 1978, while for logistic support of Argentina's Antarctic bases, I Escuadron Antartico operates from Rio Gallegos, Santa Cruz, equipped with a number of different aircraft including one Sikorsky S-61R, three DHC Beavers, three DHC Otters and one Douglas LC-47. Helicopters in FAA use total two S-61NRs, four Bell 47Gs, four Bell UH-1Ds, 14 Hughes 500Ms, eight Bell 212s delivered in 1978 for general purpose duties, and six UH-19s; three Boeing CH-47C Chinook medium-lift helicopters were ordered in 1978 but their delivery has been halted by the USA.

Training duties come under the command of the Grupo Aerea Escuela, and flying training is conducted at Cordoba air base on about 35 Beech T-34As (due for imminent replacement by a more modern type) and 12 MS760 Paris. For operational conversion, pilots move to squadrons at Brigade level.

Title: Comando de Aviacion Naval Argentina
Headquarters: Buenos Aires
Strength: 130+ aircraft

Combat

The Argentine Navy operates a single light fleet carrier, the 16,000 ton ex-Royal Netherlands Navy *Karel Doorman* (ex-HMS *Venerable*) bought in 1969, renamed *25 de Mayo* and currently the flagship of the Navy. She is equipped with an angled deck, mirror landing aid and a recently-installed computer-assisted action information system for the control of aircraft away from the carrier. Deployed aboard the ship are detachments of McDonnell Douglas A-4Q Skyhawk fighter-bombers and Grumman S-2E Tracker reconnaissance aircraft from establishments of 15 and four aircraft respectively. Both these types are shore-based at Comandante Espora Naval Air Base. Long-range maritime reconnaissance is the task of ten Lockheed SP-2H Neptunes, six being bought from the US Navy in 1966 and a further four from the same source in 1977. Also in service are a few PBY-5A Catalinas (three Grumman HU-16B Albatross being retired in 1977), long-overdue for retirement and still used for SAR and supply flights.

Transport and Support

The naval air support fleet is composed of some 20 aircraft of seven different types. For long-range transport flights No 3 Naval Air Wing employs three Lockheed L-188 Electras, three Douglas C-54s and two DC-4s from the base at Ezeiza. Short-range duties are done by eight Douglas C-47s and there is a single FMA IA50 Guarani II for VIP/transport use. One HS125-400 is based at Ezeiza for calibration work. For various liaison and communications duties there are three DHC Beavers and four Beech Queen Airs operating from Comandante Espora NAB while from Ezeiza two Beech Super King Air 200s have replaced a few obsolete Beech C-45s on liaison duties. A single DHC Twin Otter is attached to Grupo Aeronavale Antartico and when not assigned to Argentina's Antarctic force, it is home-based at Punta de Indio NAB. The helicopter force is grouped into No 4 Naval Air Wing at Comandante Espora and comprises nine Aérospatiale Alouette IIIs delivered between 1969 and 1978, three Bell 47Gs, five Sikorsky S-55s, four Sikorsky S-61D anti-submarine helicopters delivered in 1972, and two Westland Lynx HAS23s ordered for shipboard use.

Aircrew training is centred at the Naval Aviation

School at Punta de Indio NAB. To modernise the equipment a batch of 15 Beech T-34C Turbo-Mentors was delivered to the School in 1978 for basic tuition, replacing 28 North American T-28 Fennecs purchased from France in 1966 and modified for both shore and carrier use. In addition there are about 12 T-6 Texans still in service. For jet conversion and continuation training, the Navy has eight Aermacchi MB326GBs delivered in 1971 and operated in association with the A-4Q Skyhawks.

A small coastguard force known as Prefectura Naval Argentina is equipped with five Short Skyvan 3M transports and six Hughes 500M helicopters.

Title: Comando de Aviacion del Ejercito
Headquarters: Buenos Aires
Strength: 69 aircraft

The Argentine Army operates a modest aviation element which became a separate force late in 1959 and currently flies both fixed-wing aircraft as well as helicopters. The main Army Aviation base is Campo de Mayo near Buenos Aires. Largest fixed-wing type in service is the Aeritalia G222 STOL transport, the first of three being delivered in March 1977 to replace Douglas C-47s. Another recent delivery was an order for four Swearingen Merlin IIIAs for liaison and VIP transport work and these join four Piper Navajos, two DHC Twin Otters, one Beech Queen Air and a King Air, and a single Rockwell Sabreliner 75 for VIP use. Five Rockwell Turbo Commander 690As have been bought from a batch being assembled in the country, mainly for civil use while for use in the Andes three Pilatus Turbo-Porters have been bought. For tactical and AOP duties there are five Cessna 207 Turbo Skywagons and for survey and mapping flights a specially modified Cessna Citation was delivered to the Service in January 1978. Fixed-wing training is performed on five Cessna T-41Ds.

Helicopters total seven Fairchild Hiller FH-1100s, seven Bell JetRangers, 20 Bell UH-1Hs, six Aerospatiale Lamas delivered in 1975, two Bell 212s and a couple of Bell 47Gs.

Top: *One of nine Aerospatiale Alouette IIIs in service with the Argentine Navy.* / Aerospatiale

Above: *Argentine Army Cessna T207.*

Australia

Title: Royal Australian Air Force
Headquarters: Canberra ACT
Strength: 313+ aircraft

Combat

Australia has a unified defence structure involving all three military arms, the overall headquarters being situated at the Ministry of Defence in Canberra. The RAAF is the single largest aviation arm and controls all its flying elements, except training, through one of two main commands, Operational Command at RAAF Penrith, New South Wales. Foremost among the combat units are Nos 1 and 6 Squadrons, based at Amberley, Qld, equipped with 21 General Dynamics F-111Cs and assigned both land and maritime long-range strike roles. A total of 24 F-111Cs was ordered in 1963 but due to technical problems with the aircraft, the RAAF did not receive them until 1973, a 10-year gap with a resulting increase in price from $146 million to $345 million. To increase the effectiveness and widen the tasks undertaken by the F-111s, four aircraft are being fitted with infra-red cameras, sensors and reconnaissance equipment in a programme planned for completion by 1980. A third squadron (No 2, also based at Amberley) operates eight BAC/Government Aircraft Factories Canberra B20s in the PR and target-towing roles, plus four Canberra T21s for training. Air defence is the task of three Dassault/GAF Mirage IIIO squadrons; one of them, No 77 Squadron, is based at Williamtown, NSW, which forms the home base of the Mirage Wing. The two other units, Nos 3 and 75, operate from Butterworth, Malaysia, as part of Australia's commitment to the five-power agreement involving Singapore, Malaysia, the UK, Australia and New Zealand. The RAAF acquired a total of 100 single-seat Mirage IIIOs and 16 two-seat IIIDOs through the 1960s from production lines set up in Australia.

The choice of a replacement for the Mirages was under active consideration in 1978. Requests for proposals had been received from European and American manufacturers, the types under evaluation comprising the General Dynamics F-16, McDonnell Douglas F-15 and F-18, Dassault Mirage 2000 and the Panavia Tornado. To reduce costs and get the most out of the eventual choice, the RAAF is having to restrict the number of aircraft to be purchased to a single squadron and the type chosen is likely to be more of a multi-role aircraft rather than a pure interceptor as originally envisaged.

Two maritime reconnaissance squadrons are operated: No 10 Squadron at Townsville flying 10 newly-delivered Lockheed P-3C Update-Two Orions replacing 12 SP-2H Neptunes; and No 11 Squadron at Edinburgh, SA, flying 10 early-model P-3B Orions delivered in 1968.

Transport and Support

Five fixed-wing squadrons and three helicopter squadrons make up the RAAF's transport force. Backbone of the fleet are two squadrons of Lockheed Hercules: No 36 Squadron, flying 12 C-130Hs delivered in 1978 and replacing a similar number of older A versions; and No 37 Squadron with 12 C-130Es delivered in 1966, both based at Richmond, NSW. On the same base are two DHC Caribou-equipped units, Nos 35 and 38 Squadrons, with a total of 23 aircraft. At Fairbairn, Canberra, is No 34 Squadron, which uses a variety for types of VIP and Governmental duties, including two BAC One-Elevens purchased in 1968, three Dassault Mystere 20s delivered in 1967 and two HS748s. The acquisition of two Boeing 707s for long-range VIP flights is being considered. There is still a sizeable force of Douglas C-47s in service, numbering some 17 aircraft and these are likely to remain in use until 1980.

The two helicopter units, based at Fairbairn and Townsville, operate in the training role (No 5 Squadron with 31 Bell UH-1Bs) and on support and SAR duties (No 9 Squadron with 16 Bell UH-1D/Hs). A third unit, No 12 Squadron at Amberley, is flying 12 Boeing CH-47C Chinooks delivered in 1974 for Army support. Not all the Chinooks are on active service as budget restrictions have forced the temporary storage of some of them.

All training comes under Support Command and primary work is performed at No 1 Flying Training School, Point Cook, on 31 New Zealand-built CT/4 Airtrainers with a further six aircraft based at the Central Flying School at East Sale. Jet training is flown on 50 Aermacchi/CAC MB326H two-seat aircraft operating at No 2 FTS, Pearce, while at CFS East Sale instructor training is performed on 15 MB326s. These jet aircraft are also regarded as a reserve strike potential in the event of a conflict and there is a strong possibility that the force will be updated to extend their useful lives into the 1990s. The School of Air Navigation was established at East

Below: *Malaysian-based Mirage IIIO interceptor of the Royal Australian Air Force.* / RAAF

Top: *Australian Navy A-4G Skyhawks of VF-805 Squadron.* / B. C. Wheeler

Above: *One of 11 GAF Mission Masters of the Australian Army.* / RAAF

Sale in 1946 and today uses eight HS748 T2s fitted out as flying classrooms for navigation and air electronics officer training. At Williamtown, NSW, is No 2 OCU which is a fighter pilot training unit flying MB326s and No 4 Flight which trains Forward Air Controllers.

Title: Royal Australian Navy
Headquarters: Canberra ACT
Strength: 65 aircraft

Combat

Flagship of the Royal Australian Navy is HMAS *Melbourne* (formerly HMS *Majestic*), a 20,000 ton light aircraft carrier bought in 1956. She is scheduled to continue in service into the 1980s equipped with an Air Group operating in the ASW, strike and ground-attack roles as well as having an SAR capability. Three squadrons make up the Group: VF-805 with McDonnell Douglas A-4G Skyhawks, VS-816 with Grumman S-2G/E Trackers and HS-817 with Westland Sea King Mk 50s. The carrier's aircraft complement totals 14 Skyhawks, six Trackers and six Sea Kings but the normal accommodation numbers around 16 aircraft and, when not embarked, the Air Group is shore-based at Nowra Naval Air Station, NSW, known as HMAS *Albatross*. A total of 16 Skyhawks were procured between 1967 and 1971 including two TA-4G trainers, while the Tracker fleet totals some 19 machines of which 16 are ex-US Navy aircraft acquired in 1977 to replace 11 lost in a disastrous hangar fire in December 1976 at Nowra. Ten Sea Kings were bought in 1975 and these replaced Wessex in the ASW role.

Transport and Support

Other squadrons based at Nowra NAS, are HS-723 equipped with four Bell UH-1Bs, two Bell 206B-1s and four Westland Wessex HAS31Bs flying in the fleet requirements, communications and SAR roles; VC-724, flying A-4G /TA-4G Skyhawks and eight MB326Hs on fixed-wing fighter ground-attack training, fleet requirements and trials duties; and VC-851, which operates two HS748s originally used for ASW training and transport work but during 1978-79 under conversion into electronic warfare trainers.

Naval pilot training is undertaken by No 1 FTS, RAAF Point Cook, on CT/4 Airtrainers followed by advanced tuition on MB326s at the RAAF base at Pearce, WA.

Title: Australian Army Aviation Corps
Headquarters: Canberra ACT
Strength: 81 aircraft

Australia's Army Aviation Corps was created in 1968 and a major base established at Oakey, Qld, in 1971. Here was accommodated the School of Army Aviation, 1st Aviation Regiment and No 5 Base Workshop Battalion. Current fixed-wing equipment comprises 17 Pilatus Turbo-Porters of 19 originally delivered in the late 1960s and forming the inventory of the 171st Air Cavalry Flight at Holsworthy and 173rd Support Squadron at Oakey. For transport and general utility work, 11 GAF Nomad Mission Masters are in service.

The helicopter element is made up of 53 Commonwealth-assembled Bell JetRangers flown in the light observation role with a number of squadrons located with major field formations in Townsville, Qld. and Sydney, NSW.

Austria

Title: Österreichische Luftstreitkräfte
Headquarters: Vienna
Strength: 160 aircraft

Combat

The Austrian Air Force is an integral part of the Federal Army and operates a variety of aircraft from western and neutral sources. An announcement is expected by 1979 on the OLk's new combat aircraft, a technical evaluation having been prepared on four types, the Dassault Mirage F1, IAI Kfir C-2, Northrop F-5E and Saab Viggen. The choice is likely to be between the last two types with many observers expecting the decision to be made in favour of the F-5E. The aircraft due to be replaced are 37 Saab 1050 fighter-bombers of 40 purchased between 1967 and 1969, flying in two wings — a Surveillance Wing of Fliegerregimenter II and a fighter-bomber wing of Fliegerregimenter III. Each wing has two squadrons of seven/eight aircraft, based at Linz and Groz, one wing acting as a training regiment.

Transport and Support

A light transport squadron, III Geschwader of Fliegerregiment I, operates two Short Skyvan 3Ms on supply work and aerial survey from Tulln, and at the end of 1976 the unit received 12 Pilatus Turbo-Porters for operation on a variety of tasks including casevac, liaison and communication. Also on strength are seven Cessna 0-1E Bird Dogs and a Saab Safir.

Due to the nature of the country, helicopters play an important role and the OLk has more than 60 in service operating under the command of Fliegerregimenter I at Linz. The machines equip three units: Hubschraubgeschwader or Helicopter Wing I, with 13 AB206As flying in the training role but with a secondary SAR commitment, and a second squadron with 12 Bell OH-58Bs delivered in 1975-76 for the AOP task; HG II with two squadrons flying 22 Alouette IIIs on SAR duties; HG III has two Sikorsky S-65C-2 heavy-lift machines delivered in 1970, and 24 AB212 twin-engined helicopters being delivered during 1979-80 replacing 22 AB204s on utility work. Located at Tulln is the basic flying training school designated Fliegerregiment II and equipped with 20 Saab 91D Safirs.

Below: *Austrian Air Force Saab 1050E.* / B. C. Wheeler

Bangladesh

Title: Bangladesh Defence Force (Air Wing)
Headquarters: Dacca
Strength: 35+ aircraft

Formerly East Pakistan, Bangladesh became an independent state with Indian assistance in 1971. Since then a number of countries have provided military aid including India, China, the Soviet Union and the United Kingdom. The front-line combat element in the Air Wing is made up of a single squadron of nine Soviet-supplied MiG-21MF fighter-bombers and two MiG-21U two-seat conversion trainers delivered in 1973 and based at Tezgaon, Dacca. Reports have stated that a batch of 36 Chinese-built MiG-19s have been received by the wing, four being destroyed in an attempted coup in 1977, and this type of combat aircraft is now understood to be the main operational component of the Air Wing.

Transport aircraft include several Antonov An-26s and a single An-24 delivered by Russia in 1973 together with at least three Mil Mi-8s from the same source. Other helicopters total four ex-Indian Air Force Alouette IIIs used for liaison duties and six Bell 212s delivered in 1977 by Bell-distributors, Heli-Orient. A modest training unit is equipped with eight Fouga Magister jet trainers refurbished by Aérospatiale of France.

Belgium

Title: Force Aérienne Belge
Headquarters: Brussels
Strength: 283 aircraft

Combat

Belgium's air force combat units form part of NATO's 2nd Allied Tactical Air Force along with the British, Dutch and German air forces based in Europe. In June 1975, the Belgian Government announced the choice of the General Dynamics F-16 fighter-bomber to replace the FAB's Lockheed Starfighter force from 1979; a total of 116 aircraft are on order of which 12 will be two-seat F-16B combat trainers. The first of 18 F-16s are due to be delivered early in 1979 from the Sabca/Fairey production line at Gosselies, and thereafter 12 in 1980, 19 in 1981, 22 in 1982, and the remainder by 1984. The Lockheed Starfighter force due for replacement comprises the 10th Wing at Kleine Brogel with two squadrons or Smaldeel, Nos 23 and 31, each with 18 F-104Gs in the fighter-bomber role; and the 1st Wing at Beauvechain with Nos 349 and 350 Squadrons, again each flying 18 F-104Gs and operating in the all-weather fighter-interceptor role. The first unit to convert to the F-16 will be No 350 Squadron, which is expected to begin transitioning towards the middle of 1979. Two-seat TF-104Gs are on the strength of each unit. A total of 100 single-seat and 12 two-seat Starfighters were delivered to the FAB between February 1963 and mid-1965 from the Sabca production line but attrition has reduced these numbers to a total in service figure of around 82 aircraft.

The Dassault Mirage force, ordered in 1968 with deliveries beginning in 1970, consists of two tacitcal wings. At Florennes is No 2 TW comprising No 2 Squadron equipped with 18 Mirage 5BAs in the fighter-bomber role and No 42 Squadron flying 18 tactical reconnaissance Mirage 5BRs. The 3rd Tactical Wing, based at Bierset, is made up of No 1 Squadron with Mirage 5BA fighter-bombers and No 8 Squadron with two-seat 5BDs, the latter operating as an OCU while retaining its combat status. A total of 106 Mirages was originally procured for the FAB, consisting of 63 5BAs, 27 5BRs and 16 5BD conversion trainers, but nearly 20 have been lost on operations.

Transport and Support

Providing the important logistic element of the FAB, is No 15 Transport and Communications Wing based at Melsbroek and comprising two squadrons; No 20 Squadron is equipped with 12 Lockheed C-130H Hercules, delivered in 1973, and assigned heavy-lift and tactical duties, while No 21 Squadron acts as a VIP and communications unit flying two ex-Sabena Boeing 727QCs, six Swearingen Merlin IIIAs, two Dassault Falcon 20s and three HS748s.

Elementary training is flown on 33 Siai-Marchetti SF260MBs at Goetsenhoven of 36 originally delivered between 1969 and 1971. For basic training there are 40 Fouga Magisters in Nos 7 and 9 Squadrons at Brustem, while advanced tuition is flown on 12 Lockheed T-33As of No11 Squadron at the same base. To replace the Magisters, the Air Force is receiving 33 Alpha Jet 1Bs beginning in early 1979 with the order due for completion by April 1980.

An SAR flight, based at Coxyde and designated No 40 Squadron, operates five Westland Sea Kings Mk 48 helicopters which were delivered in December 1976, replacing 11 Sikorsky S-58s; one of the Sea Kings has been fitted with a VIP interior for use by the Belgian Royal family but can be re-converted back to the SAR configuration in an emergency.

Below: *Dassault Mirage 5BA of No 2 Squadron, Belgian Air Force.* / B. C. Wheeler

Top: *Belgian Navy Alouette III, one of three in use.* / Belgian MoD

Above: *Twelve B-N Defenders were delivered to the Belgian Army.* / Britten-Norman

Title: Force Navale Belge
Headquarters: Brussels
Strength: 4 helicopters

The Belgian Navy operates a single Sikorsky-Sud HSS-1/S-58 and three Aérospatiale Alouette IIIs, the latter delivered in 1970-71. The four machines are used for SAR and support duties co-operating closely with No 40 Squadron, FAB, at the coastal air base at Coxyde.

Title: Force Terrestre Belge
Headquarters: Brasschaat
Strength: 84 aircraft

The Belgian Army operates four light aircraft squadrons, largely helicopter-equipped, for AOP, liaison and communications duties. Based at Brasschaat is the Light Aviation School (No 15 Squadron), equipped with seven Britten-Norman Defender fixed-wing light transports and 24 Aérospatiale Alouette IIs (Artouste and Astazou-powered) of a total of 66 helicopters of this type in service. Nos 16, 17 and 18 Squadrons are based in Germany at Butzweilerhof, Werl and Aachen respectively, operating with NATO units, No 16 Squadron flying five B-N Defenders and 14 Alouette IIs, and the other two units having approximately 14 Alouettes each. The Defenders, based at Brasschaat and with No 16 Squadron are used for liaison and instrument training as well as for transport duties. Six Piper Super Cubs remain in the inventory for glider towing.

Maintained by the Army Air Corps and based at Brasschaat, is the small airborne arm of the Belgian police known as the Gendarmerie or Rijkswacht. It has an offical establishment of eight helicopters comprising three Aérospatiale SA330C Pumas received in 1973, and five Alouette IIs.

Benin

Title: Force Armées Populaire
Headquarters: Cotonou
Strength: 10 aircraft

This ex-French dependency was known as Dahomey before adopting its present name in 1976. A modest air arm operates mainly in the transport and liaison roles linking outlying areas in the north of the country. Three Douglas C-47s, a Cessna Reims Super Skymaster and two or three MH1521M Broussards are in use together with a single Aero Commander 500B for VIP work, plus a Bell 47G and an Alouette II — almost all supplied from ex-French military stocks. A single F27 Mk 600 was acquired in 1978.

Bolivia

Title: Fuerza Aerea Boliviana
Headquarters: La Paz
Strength: 140+ aircraft

Combat

Established as a separate military force from the Army in September 1957, the Bolivian Air Force (FAB) is steadily modernising both its front and second-line units after many years equipped with outdated Second World War aircraft. The highest-performance type at present in service is the North American F-86F Sabre, four of which were purchased from Venezuela and currently equip an Air Pursuit Group at Santa Cruz, but the Service is known to be keen to procure a more efficient interceptor and types under evaluation include the Northrop F-5E, Sepecat Jaguar and the Brazilian Embraer Xavante. Another Air Pursuit Group or Grupo Aereo de Caça, operates from El Alto airbase equipped with 15 Lockheed/Canadair T-33A/N Silver Stars bought from Canada in two batches in 1973 and 1977, and employed in the ground-attack/advanced trainer/fighter roles.

Based at Tarija is Air Cover Group No 1 (Grupo Aereo de Cobertura) flying 12 North American AT-6 Texan strike aircraft, while at Cochabamba there is a Combined Air Group (Grupo Aereo Misto) equipped with about six Bell UH-1Hs delivered in 1975, some Aérospatiale Lama rescue helicopters and a few AT-6 Texans. A Special Operations Group (Grupo de Operacoes Aerea Especiais) flies from Reboré using 12 Hughes 500M helicopters in the gunship role; these were delivered to the FAB in 1968 and not all may be in use.

Transport and Support

A military airline, Transportes Aereos Militares, forms the transport wing of the Bolivian air force, providing passenger and cargo services to remote areas of the country. Main base is El Alto, La Paz, and the largest type in use at present is the Lockheed C-130H Hercules, of which three are in service. Other aircraft flown by the unit total three remaining Convair CV-440s of six purchased in 1972, four Convair CV-580s, seven Douglas C-47s, two Douglas DC-4s, two DC-6s, six of seven Israeli-supplied IAI Aravas and one Lockheed Electra. For Presidential use the FAB has a Rockwell Sabreliner 60.

Liaison and utility flying is the task of 15 Cessna 185s, two Cessna 402s and two Turbo Centurions, a single Fairchild Turbo-Porter, two Beech Super King Air 200s and a Cessna 421B. The Military Aviation College at Santa Cruz conducts primary and advanced training on Brazilian and American aircraft types. Due however to congestion at the joint civil and military airfield, the college is to move to Santa Rosa by 1982. Present equipment is divided into two flight squadrons. The first operating about 12 Aerotec T-23 Uirapuru of 18 originally delivered in 1974 for primary cadet training, and the second unit having ten North American T-6s and two T-28 Trojans for basic work; six Italian Siai-Marchetti SF260Ms have been acquired recently to modernise this part of the training syllabus. Replacing the older piston-engined aircraft over the next couple of years will be a batch of Swiss Pilatus PC-7 Turbo-Trainers ordered in 1977 and understood to total 14. Six Cessna T-41Ds and some Cessna 310s are also in use.

The National Photogrammetry Service has at its disposal a single Gates Learjet 25B received in late 1975 and fitted with a Wild RC-10 camera; more recently a Cessna 402 joined the Service, similarly adapted for aerial photography.

Below: *Presidential Rockwell Sabreliner of the Bolivian Air Force.*

Botswana

Title: Botswana Defence Force
Headquarters: Gaberone Village
Strength: 5 aircraft

With increasing guerilla activity along the border with white Rhodesia, this small country has established a combined military unit for border patrol, casevac, communications and forward air control. Initial equipment for the airborne element of the BDF consists of two of three Britten-Norman Defender patrol aircraft ordered in 1977, the third aircraft being lost on operations early the following year. The aircraft are based at Gaberone Village and have provision for Sura rockets, Skyshout and extra fuel tankage. An order for three more Defenders was placed in 1978 together with a Botswana Government purchase of two Short Skyvan 3Ms for delivery in 1979.

Below: *Botswana Defence Force B-N Defender.* / Britten-Norman

Brazil

Title: Forca Aerea Brasileira
Headquarters: Rio de Janeiro
Strength: 840+ aircraft

Combat

The largest air arm in South America, the FAB is also one of the best equipped, having recently undergone a modernisation programme. For pure interception duties, Brazil has a single squadron of 15 Dassault Mirages purchased from France in 1970 and currently based at Anapolis, near the capital Brasilia. Designated 1st Air Defence Wing or Ala de Defesa Aérea (ALADA), the aircraft comprise 11 IIIEBRs plus four more ordered to make up for attrition, and four two-seat IIIDBR; they form an important element of a French-installed computerised air-defence system known as DACTA I and II, which by the end of the 1980s will embrace the whole of Brazilian airspace. At Santa Cruz is the 1st Fighter Group (1st Grupo de Aviacão de Caca) made up of two squadrons or Esquadrãos, flying 36 Northrop F-5Es and six F-5B two-seat trainers, three of which have been lost, ordered in 1974, deliveries were completed in March 1976.

Also assigned the fighter role are subsonic Embraer AT-26 Xavantes (licence-built Italian MB326s), which equip two squadrons of the 4th Fighter Group at Fortaleza, and a single squadron of the 14th FG based at Canoas. Xavantes perform a number of roles with the FAB, including ground-attack and training. Following the initial order for 112 aircraft placed in 1969, a further contract for 45 was received by Embraer followed by a second re-order for 10 aircraft which takes the Air Force total to 167; the extra AT-26s should all be delivered by mid-1979. Reconnaissance and attack squadrons total five, three of which have Xavantes (Nos 3, 4 and 5) while the other two fly the indigenous armed version of the Neiva T-25 Universal (Nos 1 and 2 Squadrons at Belem and Recife respectively). Complementing these units are three liaison and observation squadrons equipped with Neiva L-42 Regentes T-25s and Bell UH-1H helicopters.

The maritime branch of the air force is known as the Comando Costeiro and operates all fixed-wing aircraft as well as supplying ASW machines for Brazil's sole aircraft carrier *Minas Gerais* (ex-HMS *Vengeance* purchased in 1957). The ship accommodates six US and Canadian-supplied Grumman S-2 Trackers from a total force of eight refurbished S-2Es acquired in 1978 and forming the equipment of the 1st Carrier-borne Aviation Group (Grupo de Aviaco Embarcada) based at Santa Cruz. A similar number of older S-2As are being relegated to the transport role flying alongside a liaison unit operated by the FAB on behalf of the Navy equipped with a few Pilatus P-3s and North American T-28s. For coastal patrol duties, the 7th Air Group at Salvador has recently received 24 Embraer EMB-111A(A)s which began replacing the unit's

Top: *Brazil operates EMB-111s for maritime patrol duties.* / EMBRAER

Above: *Westland Lynx of the Brazilian Navy.* / Westland

seven long-serving Neptunes during 1978; they are designated P-95 in FAB service. For long-range rescue work, the 6th Air Group at Recife flies three Lockheed RC-130E Hercules. Two further SAR units are based at Florionapolis; No 2 Squadron equipped with 13 Grumman SA-16A/B Albatross amphibians, and No 3 Squadron with six Bell SH-1Ds and a few Bell 47Gs.

Transport and Support

Airlift capacity within the Forca Aerea Brasileira is available from four units divided into the following: No 1 Squadron, I Group, equipped with seven Lockheed C-130E, three C-130H and two KC-130H Hercules, the last-mentioned tasked with air-refuelling the F-5s; No 1 Squadron, II Group, with six British Aerospace HS748s and some EMB-110 Bandeirante light transports; No 2 Squadron, II Group, with six HS748 Srs 2Cs delivered in 1976; and No 2 Squadron of a Parachute Transport Group, equipped with some of the 21 surviving DHC Buffaloes bought by the FAB in the late 1960s. Most of the above units are stationed at the main transport base at Galeao, Rio de Janeiro, while the two squadrons of the Special Transport Group or Grupo de Transporte Especial, operate from the capital Brasilia equippped with two Boeing 737-200s, one BAC Viscount, eight British Aerospace HS125s, six Bell 206 JetRangers and some Bandeirantes. This VIP and Government transport unit more recently took delivery of five Embraer EMB-121 Xingu light twins. Twelve licence-built EMB-810C Senecas are attached to a number of units for liaison duties. Four further squadrons operate Bandeirantes and two more fly the remaining Buffaloes not flown by the Parachute Group. A total of 86 Embraer EMB-110 Bandeirantes were ordered by the Air Force made up 56 C-95 (FAB designation) light transports, 20 C-95A freighters, six R-95 photo-survey aircraft and four EC-95 navaid calibration versions. Still flying with the FAB are six Consolidated PBY-5A Catalinas which are based at Belem and used for transport and occasional SAR duties.

Helicopters equip a counter-insurgency unit at Santos and total six Bell UH-1Ds, four Bell 206s and four Hughes OH-6As. Two HS125s and the four EC-95s form a navaid checking and survey unit, while liaison and observation duties are performed by 40 Neiva L-42 Regentes and 70 U-42 Regentes respectively.

Training equipment is mainly home-produced — 100 T-23 Uirapurus for primary work, 132 T-25 Universals for basic training plus a re-order for eight to balance attrition, and some AT-26s for advanced jet training. Helicopter tuition is performed at Santos, Sao Paulo, on some 30 Bell H-13Js.

Title: Forca Aeronaval
Headquarters: San Pedro de Aldeia
Strength: 44 helicopters

The Brazilian Naval air arm has been a helicopter operator since 1965, when the President decreed that all fixed-wing aircraft were to be operated under FAB control. Consequently the Navy supplies the Sikorsky SH-3D Sea King helicopters for operation aboard the carrier *Minas Gerais*, while the air force contributes the S-2Es, although the fixed-wing type comes under naval control once aboard. Six SH-3Ds are in service with No 1 ASW Squadron and when ashore are based at the main airfield of San Pedro de Aldeia.

A total of nine Westland Lynx were ordered in 1975 for operation from six Niteroi class frigates built or under construction in the UK and Brazil, and deliveries began in 1978. For liaison, general duties and pilot training there is a support squadron equipped with three Westland Whirlwind 3s, 18 Bell 206B JetRangers and eight Westland Wasp HAS1s; six of the Wasps were received in 1977-78.

Brunei

Title: Royal Brunei Malay Regiment, Air Wing
Headquarters: Berakas Camp, Brunei Town
Strength: 11 aircraft

Formed in 1966 to support the Brunei Army and commanded by a seconded RAF officer, the Air Wing has standardised on types over the past few years and is now almost completely a helicopter force. The largest type in use is the twin-turbine Bell 212, four of which are operated on heavy-lift duties although one has a special VIP interior for the personal use of the Sultan. Three Bell 205As and three JetRangers complete the rotary-wing fleet, the latter type being fitted with uprated engines to Model 206B standard for better tropical performance. The only fixed-wing aircraft is a single HS748 which was delivered in 1971 for VIP and government flights.

Bulgaria

Title: Bulgarian People's Air Force
Headquarters: Sofia
Strength: 270 combat aircraft

With the country a member of the Warsaw Pact, the air force operates as an integral part of the Eastern Bloc alliance, fielding a mainly ground-attack force. Four interceptor squadrons form a regiment equipped with some 60 MiG-21s, while a further regiment made up of three squadrons has 36 MiG-19s. For fighter-bomber duties two regiments with six squadrons fly more than 70 MiG-17s, and a further MiG-17-equipped regiment operates in the fighter-reconnaissance role.

Transports number approximately 30 and comprise Antonov An-24s, Ilyushin Il-14s, four Il-18 government aircraft and at least one Tu-134. A helicopter regiment has 40 Mil Mi-4s and 30 Mi-8s together with a few Mi-2 liaison machines. Trainers include MiG-15UTIs and Czech L-29 Delfins, the latter expected to be replaced by the more modern L-39 Albatross. The Bulgarian Navy has about six Mi-4s and some Mi-2 helicopters for coastal patrol and SAR duties.

Burma

Title: Union of Burma Air Force
Headquarters: Rangoon
Strength: 130+ aircraft

Combat

Main task of this small air arm over the past few years has been internal security and counter-insurgency work. Operations have been conducted for some years against Communist guerrillas in the north of the country. Modern equipment remains scarce in the UBAF, due to a relatively low annual defence budget, the main combat potential being vested in six armed Lockheed AT-33As, survivors of more than a dozen supplied by the USA in the late 1960s. Complementing these are ten Italian Siai-Marchetti SF260MBs delivered in 1976 and flying in the dual strike/trainer role. In 1977, an order for 18 Pilatus PC-7 Turbo-Trainers was placed with the Swiss company and when these are delivered through 1979, it is likely that they will form a further strike element as well as flying in the training role.

Transport and Support

Transport aircraft include about six Douglas C-47s, one Fokker F27 Mk 100, four Pilatus Turbo-Porters and three Porters, plus four ex-Allegheny Airlines

Fairchild Hiller F27s. Helicopters include 12 Kaman HH-43B Huskies, 13 Alouette IIIs and 13 Kawasaki-Bell 47Gs. In 1975, 18 Bell UH-1s and a number of Bell 205s were supplied to Burma for anti-narcotics patrols but are now integrated in the UBAF. For liaison there are about 10 Cessna 180s while for advanced training the service has 12 Cessna T-37Cs.

Above: *Burmese Air Force SF260MB trainer.*

Cameroun

Title: L'Armée de l'Air du Cameroun
Headquarters: Douala
Strength: 27 aircraft

This ex-French colony has been receiving military and economic aid from France since independence in 1960, but in recent years has acquired equipment for its armed forces from other western countries. From the USA, the first of two Lockheed C-130H Hercules was delivered in September 1977, while from Britain two HS748 transports were bought, one being equipped for Presidential use and based at Douala. For internal counter-insurgency work and training flights there are four ex-French air force Fouga Magisters which arrived in 1976. Other transport types include two DHC Caribous, four Douglas C-47s, a Dornier Do28, a Beech Queen Air and seven MH1521M Broussards. Helicopters include an SA330 Puma for VIP use, two Alouette IIs and an Alouette III.

Below: *Cameroun Air Force C-130H Hercules in striking white-blue livery.* / H. Holmes

Canada

Title: Canadian Armed Forces-Air
Headquarters: Winnipeg
Strength: 680+ aircraft

Combat

Canada has one of the few unified armed forces in the world, having four commands, a Canadian Forces

Above: *Canadian CF-104G Starfighter in Tiger Meet markings of yellow and black.* / B. C. Wheeler

Europe HQ, a Northern Region HQ and a Forces Training System. Air Command, which was officially established on 2 September 1975, encompasses four operational groups (Maritime Air Group, Air Defence Group, 10 Tactical Air Group and Air Transport Group and exercises command and control over the Air Training Schools and the Air Reserve. It is responsible also for providing trained air and ground crews to operate the three Lockheed CF-104 Starfighter squadrons of No 1 Canadian Air Group, which is under the command and control of the Commander, Canadian Forces Europe.

Maritime Air Group operates four long range patrol squadrons (Nos 404 and 405 at Greenwood, No 407 at Comox, BC, and No 415 at Summerside, PEI) and a Maritime Patrol Evaluation Unit flying 26 Canadair CP-107 Argus patrol bombers. A replacement for these long-serving machines has been ordered in the form of 18 Lockheed CP-140 Auroras, a version of the Lockheed Orion. The first Aurora is scheduled to be delivered in May 1980 and the last in March 1981. Provisional plans call for No 407 Squadron to receive the first four aircraft. A total of 16 Grumman CP-121 Trackers are shore-based and employed in short-range coastal patrols with No 880 MR Squadron at Shearwater and No 406 Operational Training Squadron on the east coast at Halifax, and VU-33 at Comox on the west coast. In addition to the fixed-wing aircraft, Maritime Air Group operates a fleet of 32 Sikorsky CH-124 Sea King helicopters. Two squadrons, HS-423 and HS-443 at Halifax, provide helicopters and crews to operate from destroyers and replenishment ships of Maritime Command.

Air Defence Group is integrated with the USAF through the NORAD joint air defence agreement and has an aircraft complement of four squadrons including an operational training unit flying 56 McDonnell CF-101B Voodoos (No 409 Squadron at Comox, No 416 Squadron at Chatham, No 425 Squadron at Bagotville and No 410 Squadron also at Bagotville). No 414 Squadron has an electronic warfare role and for this task operates a few remaining Canadair CF-100 Mk 5 Canucks and some Lockheed/Canadair CT-133 Silver Stars from North Bay. Also flown by the squadron are three Dassault CC-117 Falcons converted into ECM trainers. Canada's New Fighter Aircraft programme aimed at replacing the Voodoos involves a decision in favour of either the General Dynamics F-16 or the McDonnell Douglas F-18A/L Hornet; a decision is expected in 1979 with the number required totalling some 130-150 aircraft.

No 10 Tactical Air Group operates all air resources engaged in close support of the army. It has two squadrons, No 433 at Bagotville and No 434 at Cold Lake, Alta, of Canadian-manufactured CF-5As totalling 24 aircraft with air-to-air refuelling capability. Approximately 25 CF-5As are in storage. The helicopter fleet of 56 Bell CH-118 Iroquois, 72 CH-136 Kiowas and seven Boeing CH-147 Chinooks supports six squadrons (Nos 408, 450, 427, 430, 422 and 403) in Canada; one (No 444) at Lahr, Germany, operates Kiowas in the liaison role. This latter unit forms part of Canada's contribution to NATO's 4th Allied Tactical Air Force made up of the 1st Canadian Air Group of three CF-104 Starfighter squadrons (Nos 421, 439 and 441) at Baden Soellingen employed in the conventional attack role. A fourth unit, No 417 Squadron based in Canada acts as an OTU and provides augmentation crews to bring the three European-based units up to wartime strength. Some 85 CF-104Gs and 22 CF-104D two-seaters are in the inventory. Two CC-132 DHC-7s have replaced Cosmopolitans in the transport and VIP roles with the German units.

Transport and Support

Air Transport Group has two squadrons, No 435 at Edmonton, Alta, and No 436 at Trenton, Ont, with 19 Lockheed CC-130E Hercules and five CC-130Hs; No 437 Squadron also at Trenton with five Boeing CC-137s of which two have been converted to the tanker role to support the CF-5s; and No 412 squadron at Upland, Ottawa, with seven Canadair Cosmopolitans and four Fan Jet Falcons flown for VIP and government use. SAR duties are performed by four units, No 413 Squadron at Summerside, No 424 Squadron at Trenton, No 440 Squadron at Edmonton and No 442 Squadron at Comox, equipped with some CH-118 Iroquois and CH-135 Twin Hueys, 10 CH-113/113A Labrador/Voyager helicopters, 14 CC-115 Buffaloes and eight CC-138 Twin Otters. An Air Reserve Group has been formed, operating four reserve wings and seven squadrons across the country on light transport and SAR duties.

These units comprise four units with CSR-123 Otters, one with CC-129 Dakotas, and one squadron each of CC-138 Twin Otters and CP-121 Trackers.

Air training is the direct responsibility of Air Command and the four training bases are situated at Winnipeg, Portage la Prairie, Manitoba; Moose Jaw, Saskatchewan, and Cold Lake, Alberta. The Air Navigation School at Winnipeg trains navigators to wings standard using four Hercules aircraft. At Portage la Prairie, No 3 Flying School provides initial selection and flying training for pilot candidates, using 25 Beech CT-134 Musketeers. Trainees are then posted to No 2 Flying School at Moose Jaw, where training to wings standard is carried out, using the Canadair CT-114 Tutor jet trainer of which some 85 are in service. The Air Command has a total of 158 Tutors in the inventory and a modification programme is underway to update these aircraft. From Moose Jaw, new pilots train further on CH-136 Kiowa helicopters, jet fighters, or multi-engined aircraft before joining operational squadrons. Fighter training is conducted by No 419 Tactical Fighter Training Squadron at Cold Lake. The course lasts five months and includes 92 hours of flying on the CF-5.

Central African Empire

Title: Force Aérienne Centrafricaine
Headquarters: Bangui
Strength: 26+ aircraft

This ex-French colony has been receiving aid from France since becoming independent in 1960. Current strength consists of 10 Aermacchi AL60 light transports, an ex-Sterling Airways Caravelle for the use of Emperor Bokassa, a Dassault Falcon 20 for VIP flights, three Douglas C-47s and an ex-French AF DC-3, one DC-4 and six MH1521M Broussards. A small helicopter element has a single Alouette II and a few Sikorsky/Sud H-34s. Two Aérospatiale Rallye 235 light aircraft were delivered to the country in 1978 and may be operated by the air force.

Chad Republic

Title: Escadrille Tchadienne
Headquarters: N'Djamena
Strength: 29 aircraft

Under a renegotiated defence treaty with France signed in 1976, the Chad Government recently requested French help in resisting Toubou rebels of the Chad National Liberation Front operating in the north of the country. Consequently, ten French air force Jaguars supported by Transall freighters arrived in Chad in April 1978, and began strike operations against rebel strongholds. The small Escadrille Tchadienne has been operating a unit with four ex-FAF Douglas A-1D Skyraiders in the ground attack role and at least one has been lost to surface-to-air missiles. Transport and supply duties are flown by three DC-4s, nine C-47s, a single Presidential Caravelle and some Noratlas aircraft. Four SA330 Puma helicopters are used for tactical troop transport work while for liaison there are two MH1521M Broussards, four Reims C337 Super Skymasters, and two Pilatus Turbo-Porters received in 1976.

Chile

Title: Fuerza Aerea de Chile
Headquarters: Santaigo
Strength: 250+ aircraft

Combat

Opposition to the present right-wing Chilean Government by the country's former main arms supplier, Britain, has prevented any equipment dealings between the two countries for some time. Chile has therefore turned to other sources such as the United States and Brazil. Organised into Air Group or Grupos Aereos, each made up of one squadron or Escuadrilla, the FAC has as its prime front-line unit Grupo 7 at Antofagasta in the north of the country. This unit operates 15 Northrop F-5Es and three two-seat F-5Fs received in 1976. Two fighter-bomber units at the same base, Grupo 8 and 9, are equipped with Hawker Siddeley Hunter FGA71s and T72 trainers delivered in batches from 1967. Of the total of 33 FGA71s and five T72s received, only 20 were serviceable in mid-1978 due to the lack of spares occasioned by the arms embargo. A further two light strike Grupos have Cessna A-37Bs, No 1 at Iquique and No 12 at Punta Arenas, with a total of 34 aircraft between them from two batches delivered in 1975 and 1977. For SAR and ASW duties, Grupo 2 at Quintero have a number of Grumman HU-16B Albatross amphibians.

Transport and Support

For heavy transport work, Grupo 10 at Los Cerrillos, Santiago, is equipped with two Lockheed C-130H Hercules as well as five Douglas DC-6As, one SA330 Puma helicopter, a Beech King Air and about 10 Douglas C-47s. Light utility duties and communications work is flown by Grupo 5 with 14

DHC Twin Otters (with a further six delivered in 1978) at Puerto Montt, and Grupo 6 at Punta Arenas, equipped with five Beech Twin Bonanzas. Helicopters are encompassed within Grupo 3 at Temuco and number six Sikorsky S-55Ts, six Hiller SL-4s and six UH-12Es, of doubtful serviceability, and about 10 Bell UH-1Hs. At Quintero, Grupo 11 operates nine Beech 99s for navigation training while for jet tuition there are 30 Cessna T-37Bs based at El Bosque alongside the basic training fleet totalling more than 50 Beech T-34As and a number of Cessna T-41s flown on primary flying duties. Los Cerrillos AB is also the home for the Servicio de Aerofotogrametrico which has two Gates Learjet 35s and a Beech King Air 100 fitted with cameras.

Left: *DHC Twin Otters of Grupo 5, Chilean Air Force.* / DHC

Below: *Chilean Navy Bell 206 JetRangers.*

Title: Armada de Chile
Headquarters: Santiago
Strength: 45 aircraft

The Chilean Navy air component operates five Grumman HU-16B Albatross amphibians on SAR and coastal patrol duties, and a land-based element with four ex-US Navy Lockheed SP-2E Neptunes flying long-range ASW and MR missions. Helicopter SAR duties are performed by some Alouette IIIs, four Bell JetRangers and 14 Bell 47Gs employed in the liaison and training tasks. For transport and fleet requirements there are three Brazilian Embraer EMB-110C Bandeirantes and a Piper Navajo, with half-a-dozen Beech T-34As flown on training work. In 1978, the Service received six maritime patrol versions of the Bandeirante, designated EMB-111A(N), and to supplement the transport element, four Casa C212 Aviocars were received from Spain.

Ejercito de Chile is the airborne arm of the Chilean Army and has both helicopters and fixed-wing aircraft for a variety of associated duties. Among the former types in use are nine Aérospatiale SA330 Puma assault machines, six Lama rescue helicopters, two Bell JetRangers and three Bell UH-1Hs. A fixed-wing element is used for liaison and utility duties operating four Cessna O-1 Bird Dogs, two Piper Cherokee Sixes and at least four Piper Navajos. Ten Neiva T-25 Universals, fitted with underwing weapon racks, were acquired in 1975 and fly in the dual strike/trainer role. The Chilean Army Aero Club has 18 Cessna Hawk XPs for training based at Santiago and there are a number of civilian light aircraft at the disposal of the force.

China

Title: Air Force of the People's Liberation Army
Headquarters: Peking
Strength: 4,500+ aircraft

Combat

Mainland China has the world's third largest air force but its equipment is obsolete and it has an urgent requirement for new combat aircraft. The British Aerospace Harrier V/STOL strike aircraft is high on the Chinese shopping list and if ordered, at least 100 are likely to be bought. The purchase of an initial batch and later licence production of the Rolls-Royce Spey 202 turbofan engine, for which an agreement was signed in 1975, is believed to be for the indigenous Shenyang F-12 fighter design. Little is known about this aircraft but it is understood to be single-engined and likely to embody a number of Western ideas and equipment with an in-service entry date post-1980.

The force is organised into air divisions, regiments and squadrons, three squadrons making one regiment and three regiments a division. Current AFPLA interceptor/fighter-bomber strength, estimated by official United States sources, stands at some 4,100 home-produced MiG-17s (known as the Shenyang F-4) and MiG-19s (Shenyang F-6) and only about 50 licence-built MiG-21F (Shenyang F-8) interceptors. The bomber force is composed of 60 Tupolev Tu-16 Badger medium bombers, a few Tu-4 Bulls (based on the Boeing B-29) and about 300 Ilyushin Il-28 Beagle light bombers. MR duties are performed by Beriev Be-6 flying boats.

Transport and Support

A substantial transport arm has about 400 aircraft comprising An-2s, Li-2s (Douglas C-47s), Il-14s and some Il-18s. In an emergency the civil aviation fleet of airliners, which includes 10 Boeing 707-320s, 38 HS Tridents, five BAC Viscounts and five Il-62s, could be used. Helicopters total some 300 Mil Mi-4s and 13 French Super Frelon assault machines. A training fleet comprises two-seat MiG-15UTIs, Yak-11s and Yak-18s.

The People's Naval Air Arm has about 100 torpedo-carrying Il-28 bombers together with some 400 MiG-15/17/19 fighter-bombers integrated into the air force air-defence system. About 50 Mi-4 helicopters are used for a variety of tasks.

Colombia

Title: Fuerza Aerea Colombiana
Headquarters: Bogota
Strength: 244+ aircraft

Combat

This air force, in line with some others in South America, purchased Dassault Mirages from France in 1970 and currently operates 18, which equip the single combat group at Germon Olana air base. This force is made up of 14 Mirage 5COA fighter-bombers, two 5CORs for tactical reconnaissance and two 5COD two-seat trainers. There is also a bomber/reconnaissance squadron equipped with eight Douglas B-26Ks and RB-26Cs of doubtful serviceability, and a further combat potential is available from 10 Lockheed AT-33As usually flown in the advanced trainer role. Purchased in 1977 were 10 Hughes 500M-D Defender multi-role helicopters and these are believed to operate in the Coin role.

Transport and Support

The transport element comprises the military airline Satena, which operates services linking the capital Bogota with distant points around the country. Its fleet consists of two Douglas C-54s, seven DC-3s and three HS748s. The air force Transport Command, is equipped with two ex-Canadian Lockheed C-130B Hercules acquired in 1971, one Fokker-VFW F28 Mk 1000 for VIP use, 10 DHC Beavers, three Douglas C-54s, six Pilatus Porter and Turbo-Porters, a few C-45s and 16 C-47s. Ten Cessna T-37Cs, six Northrop T-38s delivered in 1978, and 30 Cessna T-41Ds are used for training at Marco Fidel Suarez airbase, Cali, and there are around 30 Beech T-34s for basic tuition.

A large helicopter force has 27 Aérospatiale Lamas for SAR duties, 16 Bell 47Gs, 12 Hughes 500Ms, six Kaman HH-43 Huskies, six Sikorsky TH-55s, six Bell UH-1Bs and four Hiller H-23s; one Bell 212 is flown for Presidential use.

Below: *Still in widespread use, the DHC Beaver; a Colombian example is shown.* / PFI

Comores Islands

Title: Comores Air Arm
Headquarters:
Strength: 10 aircraft

This small archipelago, formerly administered by France, is situated off the coast of East Africa and established a nominal air force in 1977. Three Siai-Marchetti SF260W Warriors were bought from Italy for the dual trainer/policing role, followed by a more recent order for five SF260Cs. There is also a single Douglas C-47 and a Cessna 402B.

Congo Republic

Title: Congo Air Force
Headquarters: Brazzaville
Strength: 20 aircraft

This ex-French colony has an air arm directed mainly at transport and communications work. Equipment has been supplied by both France and the Soviet Union and includes three Douglas C-47s and three MH1521M Broussards from surplus French military stocks, a single Nord Fregate, five Antonov An-24RVs and five Ilyushin Il-14s. An F28 Mk 1000 was acquired in 1975 for VIP flights, while helicopters number three Aérospatiale Alouette IIs and an Alouette III.

Below: *One of three Soviet-supplied An-24s of the Congo Air Force.* / H. Holmes

Cuba

Title: Fuerza Aerea Revolucionaria
Headquarters: Havana
Strength: 350+ aircraft

Combat

Since the revolution in 1958, the FAR has been supplied with aircraft and equipment from the Soviet Union and in return has contributed pilots and personnel to a number of military actions abroad, notably in Africa. Current combat strength comprises five interceptor squadrons flying 50 early-version MiG-21Fs and 30 later MiG-21MFs. Some MiG-23s, possibly as many as 20, are reported, to be in service, and Soviet Air Force pilots on exchange posting are regularly flying patrols in these aircraft as well as in the MiG-21s. Supplementing these units are two squadrons with 40 MiG-19s and four fighter-bomber squadrons equipped with 75 MiG-17s. A second-line training unit has 15 MiG-15s and some MiG-15UTI two-seaters. Soviet long-range reconnaissance bombers regularly use Cuban bases for refuelling.

Transport and Support

For transport duties there are about 50 aircraft made up of An-2 biplanes, Il-14s and An-24s, while the helicopter element has approximately 30 Mil Mi-1s and 24 Mi-4s. Training is conducted in Cuba and the Soviet Union, primary tuition in the country being flown on Zlin 326s followed by jet conversion on MiG-15UTIs and MiG-21Us.

Czechoslovakia

Title: Ceskoslovenske Letectvo
Headquarters: Prague
Strength: 800+ aircraft

Combat

A member of the Warsaw Pact, Czechoslovakia contributes a mainly tactical air force to the organisation. Made up of air regiments, each having up to three squadrons, the air force has six regiments with 18 squadrons flying 170 MiG-21s and 80 MiG-19s, four regiments with 12 ground-support squadrons operating 70 Sukhoi Su-7s and about 100 MiG-17s, and two regiments with six reconnaissance squadrons equipped with 80 MiG-21Rs. A recent acquisition in 1978 is a squadron of 12 MiG-23 variable-geometry interceptors based at Pardubice with the liklihood that the unit will expand to full regiment status in the future.

Transport and Support

Approximately 50 transport aircraft are in use, mainly Il-14s with an increasing number of An-24s and An-26s replacing them; a single Tu-134 is operational for VIP and Government flights. Helicopters include some 100 Mil Mi-4s and Mi-8s while the training fleet includes 150 L-29s (with the new L-39 Albatross replacing them on a one-for-one basis), MiG-15UTIs, MiG-21Us, Yak-11s, and Zlin 326s.

Below: *MiG-21F day fighter of the Czech Air Force.*

Denmark

Title: Flyvevaabnet
Headquarters: Vedbaek air station
Strength: 205 aircraft

Combat

Assigned to NATO's Allied Air Forces Northern Europe in a conventional tactical role, the Royal Danish Air Force has two main operational commands, Air Tactical and Air Material Command. Under the former organisation, known as Tacden (Tactical Air Commander Denmark), a total of 51 Saab 35XD Drakens are allocated to No 725 Squadron at Karup, with 20 F35XD fighter-bombers and three TF35XD two-seaters; No 729 Squadron at the same base, operates 20 RF35XDs and three TF35XDs in a reconnaissance role. Two squadrons of North American F-100D/F Super Sabres (Nos 727 and 730 at Skrydstrup) also fly in a tactical role with approximately 24 single-seaters and 14 two-seat F-100Fs; the majority were acquired from ex-USAF stocks with initial deliveries made in 1959 of 48 single-seaters and 24 two-seaters. Attrition has reduced the force to the present figures and these aircraft have undergone a refurbishing programme to extend their service lives into the early 1980s when they will be replaced by General Dynamics F-16s. A total of 46 single-seat F-16As and 12 two-seat F-16Bs are on order for the RDAF with the first squadron, No 727, forming towards the end of 1979, followed by No 730.

A total of 47 Lockheed F-104G Starfighters are in the RDAF inventory, including 15 ex-Canadian AF CF-104Gs and seven two-seat CF-104Ds. Two interceptor units operate these machines, Nos 723 and 726 Squadrons at Aalborg, with some of the Ds flying in the ECM role. Original Starfighter purchase occurred in 1964, the Canadian aircraft arriving in 1971.

Above: *Due for replacement by F-16s are the Danish F-100 Super Sabres.* / B. C. Wheeler

Transport and Support

The main RDAF transport base is Vaerlose, north-west of Copenhagen, where No 721 Squadron operates three Lockheed C-130H Hercules delivered from 1975. At the same base, No 772 Squadron has an SAR role for which it is equipped with eight Sikorsky S-61As. Detachments of these helicopters are also based at Aalborg and Skrydstrup. Training is performed at Avno on Saab T-17 Supporters, final delivery of the 32 aircraft being made in 1977 of which 21 are assigned the air force primary training role and the remaining nine operating with the Danish Army in the observation role. After the primary stage, student pilots go to the USA for a course on USAF Cessna T-41s, progressing to Cessna T-37s and T-38s before returning to Denmark for conversion to the two-seat TF-100F, TF-104G and TF35.

The Army Flying Service is headquartered at Vandel and flies in support of Danish army ground forces. Observation duties are performed by the nine Saab Supporters and 15 Hughes 500M light helicopters delivered in 1971-73.

The Danish Navy has a small air component equipped with eight Aérospatiale Alouette IIIs for liaison duties; they are maintained at Vaerlose by the RDAF's No 722 Squadron. In 1977, the Navy ordered seven Westland Lynx for delivery in 1979. They will be used from frigates and shore bases on reconnaissance and fishery protection in Greenland, the Faroes and the North Sea.

Dominican Republic

Title: Fuerza Aerea Dominicana
Headquarters: Santo Domingo
Strength: 65+ aircraft

This small air force operates under a central air command within which are a number of squadrons operating a variety of relatively obsolescent aircraft. The fighter squadron, equipped with more than a dozen North American F-51D Mustangs, acts as a training unit with a secondary combat role. About seven North American B-26K Invaders, of doubtful serviceability, form a bomber component, and the survivors of 10 de Havilland Vampire F1/FB50s plus two Lockheed T-33As constitute the country's sole jet equipment. The Vampires are ex-Swedish acquired in the mid-1950s and have a fighter-bomber role. For Coin duties approximately six North American T-28Ds are flown, but these machines also perform training tasks. With the country almost surrounded by water, the SAR and MR roles assume considerable importance but consistently low defence budgets have seen no replacement for the two Consolidated PBY-5A Catalinas used for some time.

A transport squadron operates six Douglas C-47s, six Curtiss C-46s, and three DHC Beavers. Helicopters include three Aérospatiale Alouette II/IIIs, two Sikorsky H-19s, two Hiller UH-12Es, and seven Hughes 500Ms. The training element has four Cessna 172s for primary duties, plus some North American T-6s and Beech T-11s.

Dubai

See under United Emirates Air Force

Ecuador

Title: Fuerza Aerea Ecuatoriana
Headquarters: Quito
Strength: 200+ aircraft

Combat

This South American air arm is undergoing an equipment modernisation programme which has involved orders for new aircraft from countries as far apart as Canada, France, Israel and the UK. The most recent purchase was for 16 Dassault Mirage F1EJ fighter-bombers and two F1BJ trainers, the order following America's veto on the sale to Ecuador of 24 Israeli Kfirs in 1977. The Mirages will complement the squadron of 12 BAC/Breguet Jaguar International strike aircraft delivered in 1977 made up of ten single-seaters and two trainers, and a bomber squadron with three BAC Canberra B6s of six originally delivered in 1954-55. Of the 16 BAC Strikemaster Mk 89s received, 14 survive equipping a strike/trainer unit. A counter-insurgency squadron was established in 1976 following the purchase of 12 Cessna A-37Bs from the USA.

Transport and Support

The Military Air Transport Group of the FAE operates four Douglas DC-6s, one of two Lockheed C-130H Hercules delivered in 1977 (the other crashing near Quito in 1978), two DHC Buffaloes delivered in 1976 and a DHC Twin Otter. There are five British Aerospace HS748s in service, two being Series 2A aircraft fitted with large freight doors. Of these machines, two operate with TAME (Transportes Aereos Nacionales Ecuatorianos), the military airline branch of the FAE. Scheduled passenger and cargo flights are flown by this airline, which also has four Lockheed Electras and a single DHC Twin Otter, all based at Guayaquil. For long-range flights, the air force makes occasional use of two Ecuatoriana Boeing 720s.

A helicopter element has two Aérospatiale SA330 Pumas, four Lamas for rescue duties in the mountains, six Alouette IIIs, and some recently-delivered Bell 212s. For navaid checking and calibration, the Air Force operates a single Beech King Air 90. A primary training fleet comprises 20 Cessna T-41s and 24 Cessna 150 Aerobats while the basic stage is catered for by 20 Beech T-34C Turbo-Mentors delivered during 1978 and replacing North American T-28s; 12 Lockheed T-33As and some of the Strikemasters fly in the advanced trainer role. Twelve Italian SF260 primary trainers were also received by the FAE in 1977.

The Ecuadorian Army or Aviacion del Ejercito Ecuatoriana has a small supporting air element equipped with a single Short Skyvan, three Pilatus Turbo-Porters and five IAI Aravas from Israel. For photo-survey use, the Army's Instituto Geografico Militar has a Gates Learjet 25D, a Beech Queen Air 80 and a Lama helicopter.

Aviacion Naval Ecuatoriana is a naval liaison unit equipped with two Alouette III helicopters, one Arava transport, one Cessna 320E, one Cessna 177 and two Cessna T-41Ds for training work.

Below: *DHC-5 Buffalo in service with the Ecuadorian Air Force.* / DHC

Egypt

Title: Arab Republic of Egypt Air Force
Headquarters: Cairo
Strength: 1,500 aircraft

Combat

After more than 20 years in the Soviet sphere of influence, Egypt is now firmly in the Western camp and urgently seeking to replace the large amount of Russian equipment in her armed forces. The air force has been particularly vulnerable as spares stocks have dwindled and the complex combat aircraft have become grounded in greater numbers. To overcome this shortage, the United States has agreed to sell Egypt 42 Northrop F-5E fighter-bombers and eight F-5F trainers. It is known that President Sadat is anxious to obtain more advanced warplanes and one type that has been constantly at the forefront of reports is the French Dassault Mirage F1 but no order has yet been placed. Instead the Egyptian Air Force has received a total of 52 Mirage IIIEE fighter-bombers, of which 38 were financed by Saudi Arabia and delivered in 1973, followed by a further order for 14 placed in 1977.

The EAF's current front-line equipment is still predominantly Russian in origin, and total numbers given reflect aircraft acquired and not necessarily those operational. A bomber force has some 25 Tu-16 Badger Gs, some armed with AS-5 Kelt ASMs, and a similar number of Il-28 Beagle light bombers. Fighter-bomber units are equipped with about 120 Sukhoi Su-7s, although some reports suggest a figure nearer 80, and nearly 150 MiG-17s, which also double in the low-level interceptor role. Also in the strike role is a MiG-27 Flogger D unit with 18 aircraft delivered in 1974, and some 48 Su-20 variable-geometry attack aircraft relinquished by the departing Russian 'advisers'.

Interception missions are flown by some nine squadrons in Air Defence Command equipped with 108 MiG-21MFs integrated into a Soviet-established air-defence system which includes SAMs and radars. There are also about 150 MiG-21PFMs and about 100 F attack aircraft; a few are equipped with under-fuselage cameras for tactical reconnaissance. In order to keep the majority of the MiG force operational, British equipment companies have received contracts to overhaul the engines and fit the aircraft with Ferranti inertial nav platforms and Smiths weapon-aiming head-up displays. China has also assisted by providing spares and about 30 engines. Supplementing the -21s are about 24 MiG-23 Flogger B interceptors and six two-seat MiG-23 Flogger C trainers supplied at the same time as the -27s.

Transport and Support

Of the six Lockheed C-130 Hercules — four C-130Hs and two EC-130Hs delivered in 1977 — one has been lost in Cyprus during an abortive commando raid in 1978 and a further 14 have been requested to replace the 25 Antonov An-12 in use. Other transports include 40 Il-14s and a VIP flight stationed at Cairo West equipped with a Boeing 707, Boeing 737 and a Dassault Falcon 20. Helicopters are operated mainly by the Army and Navy and include a batch of Westland Commando assault aircraft comprising five Mk 1s and 23 Mk 2s, including two VIP machines. From France, a further 12 Aérospatiale SA342 Gazelles have been received to add to the 42 already in use, some being equipped with HOT missiles for anti-tank use and others flying in the AOP and electronics roles. Russian types include 70 Mil Mi-8s, many armed for assault duties, 20 Mi-4s, and about 12 Mi-6 heavy-lift machines. In January 1978, a major order for 250 Westland Lynx utility helicopters was announced, the first 20 machines being built in Britain with deliveries due to begin in January 1980, followed by the assembly of knocked-down components in Egypt and finally full scale production in the country for both the Egyptian armed forces as well as for other Arab states. The Navy, as well as operating some Gazelles, also has a squadron of six Westland Sea King Mk 47s for ASW duties. Six Fournier RF4 motor gliders are flown for AOP and Elint sorties.

Training is performed on 100 Czech L-29 Delfins, some fitted with weapon pylons and equipment for the ground-attack role, and about 50 two-seat MiG-15UTIs, Su-7Us and MiG-21Us. Navigation training is done on converted Il-14 transports while basic tuition is flown on the survivors of 90 Al Kahiras (Hispano HA-200s); there are still some 200 Gomhouria elementary trainers in use, a version of the wartime Bucker Bu181. For liaison duties there are more than a dozen Polish Wilgas in service. In September 1978, the Arab Organisation for Industrialisation signed a provisional contract for up to 160 Alpha Jet trainers for delivery in the 1980s.

Below: *Egyptian Tu-16 Badger bomber.*

Ethiopia

Title: Ethiopian Air Force
Headquarters: Addis Ababa
Strength: 160+ aircraft

Combat
Following Russian support for Ethiopia during the Ogaden war with Somalia in 1977-78, the EAF front-line inventory has almost doubled from a mixture of Western types to a heavy bias of Soviet equipment. Most observers put the number of MiG-21s in service at about 50 and there were reports early in 1978 of MiG-23 variable-geometry strike aircraft in action. Cuban and Soviet Air Force personnel operate most of the newly-acquired equipment, Ethiopians generally remaining with the surviving Western types. These consist of two BAC Canberra B2 bombers of four bought in 1969, together with about 12 Northrop F-5As and two F-5B two-seaters. Until recently the air force had a squadron of 11 North American F-86F Sabres in service but their present status is likely to be less than operational. A light strike unit has six North American T-28Ds based at Dire-Dawa.

Transport and Support
The transport force has some 29 aircraft made up of 12 Douglas C-47s, 12 Fairchild C-119K Packets, two Douglas C-54s, two de Havilland Doves and an Il-14 donated by Russia some years ago. However, the recent influx of Soviet equipment into the Service could find a number of An-24/26 aircraft in use. Helicopters total five Aérospatiale Alouette IIIs, one SA330 Puma, two Mil Mi-8s and six Agusta-Bell AB204Bs; during the war Mil helicopters conducted assault operations and it is likely that the Mi-8 force has been considerably increased. Trainers include 11 Lockheed T-33As, 20 Saab Safirs and some T-28s, plus a few two-seat MiG-21Us.

The Ethiopian Army has a small support force equipped with six Bell UH-1Hs, four DHC Otters and three DHC Twin Otters.

Below: *Second of a batch of three DHC Twin Otters for the Ethiopian Army.* / DHC

Finland

Title: Ilmavoimat
Headquarters: Helsinki
Strength: 154 aircraft

Combat
Under the terms of the 1947 Treaty of Paris, the Finnish Air Force is limited to a strength of 60 combat aircraft and 3,000 personnel. The air arm is currently organised into three Air Defence Wings, each assigned to one of three air-defence areas into which Finland is divided. The main wing bases are at Rovaniemi, Tampere and Kuopio-Rissala, each with one front-line squadron together with the necessary support elements. The aerial defence of Northern Finland is the task of the Lapland Wing, made up of No 11 Squadron at Rovaniemi flying 27 Saab Drakens on all-weather fighter-bomber duties. The Draken force is made up of 12 J35Ss recently assembled by the Finnish company Valmet at Halli; six ex-Swedish AF J35BS aircraft orginally leased to the Finns for training in 1973 and subsequently purchased by them; and a further nine low-houred ex-Swedish AF machines comprising three two-seat S35C combat trainers and six J35F interceptors.

Covering southern Finland at Pori is the Satakunta Wing with No 21 Squadron, equipped with armed Fouga Magisters for advanced and tactical training, while at Kuopio-Rissala is the Kerelia Wing, forming

Above: *Saab Drakens equip No 11 Squadron of the Finnish air arm.* / Kivikero

the defence of South-east Finland. The main unit in the wing is No 31 Squadron flying 18 Soviet-supplied MiG-21F day fighters and a conversion flight with three surviving MiG-21U two-seaters of four originally supplied. A replacement for these ageing aircraft has been selected, 30 late-model MiG-21bis interceptors having been ordered in 1978 from Russia, the first arriving in October of that year. At least one two-seater combat trainer version of the MiG-21bis has been received.

Transport and Support

A transport squadron is based at Utti and operates both fixed-wing and helicopter types. Eight Douglas C-47s form the backbone of the unit, supported by six Mil Mi-8s, three Mi-4s, one Agusta-Bell AB206 JetRanger and one Hughes 500C. Two Cessna 402s and five Piper Cherokee Arrows are on lease to the air force for liaison duties from Tikkakoski.

Providing a secondary combat potential but generally flown in the advanced trainer role are 50-60 Valmet-built Magisters of 80 originally acquired. Late in 1976 it was announced that 50 British Aerospace Hawk Mk 51s had been ordered to replace the Magisters, the first four aircraft due for delivery in 1980 and the remaining 46 being assembled under licence by Valmet. Selected in the same year was the indigenous Vinka primary trainer intended to replace 25 Saab Safirs at the Central Flying School at Kauhava. A contract for 30 aircraft was signed with deliveries due to begin in April 1979. Advanced conversion training is performed by No 31 Squadron on the two-seat MiG-21Us and by No 11 Squadron on the two-seat Drakens.

France

Title: L'Armée de l'Air
Headquarters: Paris
Strength: 1,500+ aircraft

Combat

France's airborne nuclear strike force of 50 Dassault Mirage IVA twin-jet bombers has been in service since 1968, when the 62nd and final aircraft was received by the service. Of the surviving machines, 36 are on call at any time, day or night, equipping six squadrons or Escadrons in two wings or Escadres (91 and 94 Escadres) dispersed among some six bases. Twelve Mirage IVAs have been converted for long-range reconnaissance at both high and low level. Supporting the bombers are three tanker squadrons, each with four Boeing KC-135Fs, the 11 remaining aircraft being based at Istres. This strategic force is expected to remain in service until 1985, when silo-based S-3 missiles will form the country's strategic deterrent until the year 2000.

The conventional air force has three main commands, Air Defence, Tactical and Air Transport, operating within four air regions (No 1 with headquarters at Metz, No 2 at Paris, No 3 at Bordeaux and No 4 at Aix-en-Provence). Air Defence Command or Commandement Air des Forces de Defence Aérienne (CAFDA), has an interceptor network linked to NATO's Nadge chain and operating within the Strida air defence system. Six squadrons are presently equipped with Dassault Mirage F1Cs (two each in the 5th Wing at Orange, the 12th Wing at Cambrai and the 30th Wing at Reims) of a planned procurement of 214 aircraft for eight squadrons in four wings; initial order was for 105 F1s followed by a further 63 aircraft of a planned second batch of 109. At Creil, No 10 Wing operates two squadrons of delta-wing Mirage IIICs and is due to convert to F1s by 1980, the 1978 defence budget having set aside funds for 33 F1s for this unit.

Tactical Command, or Commandement Aérien Tactique, is divided into 1st and 2nd CATac and controls seven squadrons of Mirage IIIE fighter-bombers with a combined conventional/nuclear attack role (No 2 Wing at Dijon with two squadrons and the Mirage OCU with IIIB/IIIC/BEs and No 4 Wing at Luxeuil with two squadrons; No 13 Wing at Colmar, which has three squadrons with Mirage IIIE and the 48 ex-Israeli 5Fs). The main tactical reconnaissance base is Strasbourg which houses No 33 Wing with its three squadrons flying nearly 50 Mirage IIIR/RDs and a few Lockheed RT-33As. SEPECAT Jaguar strike aircraft, of which the last of 200 will join the force in 1979, equip a total of nine squadrons in three wings — Nos 3 and 7 Wings at Nancy each with three squadrons and No 11 Wing at Toul-Rosieres with three units. In

addition, the 92nd Wing at Bordeaux, previously flying Vautors in two squadrons is re-equipping with 20 Jaguars. At Cazaux is No 8 Wing which specialises in weapon training and jet conversion flying using Mystere IVAs but due to convert to Alpha Jet Els in 1982; a total of 200 of these twin-jet trainers are due to be delivered to the French air force, the first arriving in November 1978. Another new and very important aircraft ordered by the Service is the delta-winged Dassault Mirage 2000, scheduled to replace the existing Mirage IIIs from 1982. The planned initial order is for 127 aircraft with total procurement expected to reach 250-300, and although specified as a multi-role aircraft, the 2000 will have a bias towards interception/air superiority but with a ground-attack capability.

Transport and Support

Air Transport Command, or Commandement du Transport Aérien Militaire (CoTAM), comprises three squadrons of C160 Transalls totalling some 46 aircraft in No 61 Wing with headquarters at Orleans, and four squadrons of Nord Noratlas totalling nearly 120 aircraft (two squadrons in No 62 Wing at Reims and two squadrons in No 64 Wing at Evreux). To replace the Noratlas, 25 new-production C160F Transalls are being built, the first machines off the new line being due in 1980 with delivery planned between 1981-83 to No 64 Wing; the aircraft in the 62nd Wing will be progressively withdrawn during the 1980s. For paratroop training and navigation work the Air Force is reportedly due to order about 12 BA HS748s.

Above: *French Air Force Dassault Mirage 5F of No 13 Wing.* / J. M. C. Guhl

Bottom: *Breguet Alize ASW aircraft of the French Navy.* / ECP Armées

For long-range VIP flights, one squadron of No 60 Wing at Roissy operates four Douglas DC-8Fs, while a second squadron at Villacoublay has a mixed fleet of VIP/liaison types including a Caravelle, five Dassault Falcons, two Aérospatiale Pumas and three Cessna 411s. For flights between France and her Pacific bases, a third squadron in No 64 Wing at Evreux has three ex-UTA airline Caravelle 11Rs. No 65 Wing has two squadrons flying short-range liaison duties with Nord Fregates, Falcon 20s, M-S Paris and Broussards. Five helicopter squadrons are concerned mainly with liaison and SAR duties and have about 50 Alouette II/IIIs and 10 SA330 Pumas. The training units will be steadily modernised over the next few years with the arrival of Alpha Jets, while twin conversion continues on 34 Dassault Flamants destined for service until at least 1980. Miscellaneous types include eight Cessna 310Ns, two 310Ks and two 310Ls. A large Magister training and liaison force totals some 300 aircraft.

Overseas, the air force has conducted strike operations with Jaguars, supported by Transalls, against Polisario guerrillas from bases in Senegal and Mauretania. Other support elements operate from St Denis Réunion in the West Indies and from Papeete in the Pacific, mostly flying Alouette helicopters and Noratlas transports. Based at Noumea is ETOM 52 equipped with three Puma helicopters for support work.

Title: Aéronautique Navale
Headquarters: Paris
Strength: 270+ aircraft

Combat
The French Navy maintains two 27,300 ton aircraft carriers, *Clemenceau* and *Foch*, with a strike force of several units. In the strike role there are two squadrons, Flotille 11F and 17F, with 36 Dassault Etendard IVMs and based at Landivisiau and Hyeres respectively, and a reconnaissance squadron, Flot 16F, based at Landivisiau, with 14 Etendard IVPs. For the interceptor role there are 32 LTV F-8E(FN) Crusaders in two squadrons based at Landivisiau, Flot 12F and 14F. A total of 71 Super Etendards are being delivered to the Navy to replace both the old Etendards and the Crusaders, the last due to arrive in mid-1981. Carrier-based fixed-wing ASW work is performed by two Breguet Alizé squadrons, Flot 4F and 6F, home-based at Lann-Bihoué and Nîmes-Garons respectively; a total of 46 aircraft remain on strength, of which 28 are being modernised with new radar and ESM, and are due for service from 1980. Flot 32F supplements the Alizé units aboard the carriers, being equipped with ten Super Frelon ASW helicopters and operating from Lanvéoc-Poulmic. Maritime Patrol Command encompasses the shore-based MR units covering the Channel, Atlantic and Mediterranean areas. Thirty-five Breguet Atlantics equip four units — Flot 21F and 22F at Nîmes-Garons and 23F and 24F at Lann-Bihoué — and 14 Lockheed SP-2H Neptunes fly with Flot 25F, also at Lann-Bihoué. An updated version of the Atlantic, known as the M4 or 'Atlantic Nouvelle Generation', is under development by Dassault-Breguet for service in 1985.

The 11,000 ton helicopter carrier *Jeanne d'Arc* is used as an officer training ship but can be operated in the ASW role with accommodation for some eight helicopters. A second vessel of 18,000 tons has been ordered, designated PH-75 and due for service inthe 1980s. Helicopter-equipped units include Flot 34F based at Lanvéoc-Poulmic, flying Alouette IIIs for shipboard detachment, Flot 31F at St Mandrier with 12 Sikorsky-Sud HSS-1s for ASW duties and due to receive Westland Lynx, Flot 33F at the same base with 16 HSS-1s flying in the commando assault role but scheduled to be replaced by five Super Frelons in 1979; Escadrille 22S and 23S Alouette III units fly communications duties and supply helicopters for planeguard work aboard the carriers. The Aéronavale Lynx order stands at 26 aircraft for ASW and surface strike, with deliveries beginning in 1978 to replace ASW Sikorsky-Sud HSS-1s and Alouettes.

Transport and Support
Second-line units include Escadrille 2S at Lann-Bihoué, flying three Piper Navajos and four Nord 262s of 21 supplied to the service; Escadrille 3S at Hyères with four 262s and five Navajo; Escadrille 20S at St Raphael, operating as a flight-test wing mainly equipped with helicopters; and Escadrille 12S, based in Papeete with three specially modified Neputnes. Fifteen M-S Rallye 100s are used for training, together with some Alizés and the shipboard version of the Fouga Magister known as Zéphyrs.

Five Dassault Falcon 10MERs are used for radar training and fleet support at Landivisiau and Hyères, while nine Morane Paris operate from the former base as high-speed transports. Overseas, a number of units operate various second-line types, including Escadrille 9S with Neptunes, Douglas C-47s and C-54s on New Caledonia and some Alouette IIIs on Hao.

Title: Aviation Légère de l'Armée de Terre (ALAT)
Headquarters: Paris
Strength: 714+ aircraft

Like many of the world's army aviation formations, France's ALAT is almost entirely helicopter-equipped and provides observation and liaison services for Army ground forces. The force is organised into Regiments d'Helicoptères de Combat (RHC), of which there will be six by 1979, each having seven escadrilles or squadrons, comprising two reconnaissance units equipped with Aérospatiale Gazelles, three anti-tank units equipped with Alouette IIIs armed with Nord SS11s and later Gazelles with HOT, and two *escadrilles de manoeuvre* equipped with SA330 Pumas.

Two of the Regiments are attached to each of the French Army's two army corps, and the command of each regiment is integrated with corps headquarters. The remaining two RHCs are held in reserve. In addition, one Groupe d'Helicoptères Légères (GHL) composed of 30 Gazelles and Alouette IIs and IIIs is attached to each of the army corps for utility work such as casevac, liaison and AOP. Another GHL of 20 light helicopters is attached to each of the six French military territorial regions. ALAT's helicopter inventory includes 140 Pumas, 220 Alouette IIs for the LOH role and 84 Alouette IIIs, many equipped with SS11 missiles for anti-tank duties. A total of 170 Gazelles are being delivered, with more than half in service, replacing the Alouette IIs. ALAT intends to retain the SS11-armed Alouette III force in service alongside the HOT-armed Gazelles for the foreseeable future.

Fixed-wing Cessna 0-1s and M-H Broussards continue in use at the ALAT basic pilot training base at Dax in south-west France, trainees proceeding to Alouette IIs and finally to Gazelles. There are also some Piper Super Cubs and Tripacers.

Gabon

Title: Force Aérienne Gabonaise
Headquarters: Libreville
Strength: 32 aircraft

Formerly part of French Equatorial Africa, Gabon gained its independence in 1960 but continues to receive economic and military aid from France. The

Above: *Gabonaise C-130H Hercules.* / H. Holmes

nation's air arm is mainly concerned with communications and transport duties, but in 1975 it announced an order for six Dassault Mirages — five 5Gs and one two-seat 5DG — for the first combat squadron. Delivery of these machines is expected in 1978. The transport element operates one Lockheed L-100-20, an L-100-30 and a C-130H Hercules, all being chiefly used for heavy engineering work within the country. A single Grumman Gulfstream II is flown for VIP use together with a Dassault Falcon 20E; two Japanese NAMC YS-11As operate on joint civil/military work, and there are also three Douglas C-47s, three Nord 262s, four MH1521M Broussards, one Reims C337 Super Skymaster, three of four SA330 Pumas delivered and four Alouette IIIs.

Germany, East

Title: Luftstreitkräfte und Luftverteidigung
Headquarters: Strausberg-Eggersdorf
Strength: 500+ aircraft

Combat

A founder-signatory of the Warsaw Pact, East Germany fields a large tactical air force and is the sole Pact country to make its force subordinate to the combined supreme command controlled from the USSR. Soviet units based in the country come under the command of the 16th Air Army, Group of Soviet Forces in Germany, and fly some 1,100 aircraft comprising MiG-21SMTs, -23s, -25s, -27s, Su-17s, Mi-8s and Mi-24s. The LSK is divided into two air-defence divisions: the 3rd Air Defence Division based at Neubrandenburg covering the north with the 2nd Fighter Regiment equipped with 36 MiG-21s and the 9th FR at Peenemunde with 36 Su-7s; and the 1st Air Defence Division based at Cottbus covering the south with the 1st FR flying 36 MiG-21s, the 7th and 8th FR also believed to be flying -21s, and the 3rd FR at Preschen with Su-7s. As with most Eastern Bloc air forces, a Regiment is made up of three squadrons with an establishment of some 12-15 aircraft. All the interceptor and fighter-bomber units are currently equipped with a total of some 280 MiG-21MF/FLs 80 Su-7s, and about 30 MiG-17s in a second-line role; more modern combat types are expected to join the LSK in the near future such as the variable-geometry Su-17 and MiG-23/27.

Transport and Support

A relatively small transport force operates in support of the combat elements and comprises more than a dozen ageing Il-14s, about 20 An-24s and some An-2s. A VIP unit has a couple of Il-18s, three Tu-124s and two Tu-134s. A helicopter regiment has some 80 Mi-4s and Mi-8s while the East German Navy operates about eight of the former on SAR duties in the Baltic. Trainers include Yak-18s and Zlin 226s for primary work. Czech L-29s (being replaced by the later L-39 Albatross) and MiG-15UTIs for advanced training, and Su-7Us and MiG-21Us for conversion at a fighter training division.

Germany, West

Title: Luftwaffe
Headquarters: Bonn
Strength: 1,200+ aircraft

Combat

The air arm of the Federal Republic of Germany forms a major component in NATO's 2nd and 4th Allied Tactical Air Forces. Main combat aircraft is the McDonnell Douglas F-4 Phantom ordered in the late 1960s, a total of 260 F-4F/RF-4Es of 273 delivered being in use. These aircraft equip two fighter interceptor Geschwader, JG71 at

Above: *Luftwaffe Fiat G91R-3s will be replaced by Alpha Jets in the 1980s.* / B. C. Wheeler

Wittmundhafen and JG74 at Neuberg, assigned to 2nd ATAF and 4th ATAF respectively. Also flying F-4Fs are two fighter-bomber Geschwader: JaBoG35 at Pferdsfeld in 4th ATAF, and JaboG36 at Rheine-Hopsten in 2nd ATAF. Completing the Phantom force are 88 RF-4Es equipping two reconnaissance Geschwader, AG51 at Bremgarten in 4th ATAF and AG52 at Leck in 2nd ATAF.

Backbone of the Luftwaffe remains the Lockheed F-104G Starfighter, some 428 surviving of 604 procured from European licence manufacture, and these equip four Geschwaders — JaboG31 at Noervenich in 2nd ATAF, JaboG32 at Lechfeld (4th ATAF), JaboG33 at Buchel (2nd-ATAF) and JaboG34 at Memmingen (4th ATAF). The Starfighter OCU at Jever flies single and two-seaters. Replacing these aircraft from 1979, are the first of 324 Panavia Tornados ordered for the Luftwaffe and Marineflieger, of which the former will receive 212 Tornados for the four Starfighter units. Another new type destined for service is the Alpha Jet 1A light attack and tactical training aircraft. The first of 175 is due to join Waffenschule 50 at Furstenfeldbruck in February 1979, replacing the present Fiat G91Ts; when re-equipped the unit will be redesignated JaboG49. Of the 240 Fiat G91R/3 attack aircraft on Luftwaffe charge, only about 100 remain fully operational, flying with two wings, LeKG41 at Husum and LeKG43 at Oldenburg, both in 2nd ATAF. When re-equipped with Alpha Jets, these two units will be redesignated fighter-bomber Geschwader, JaboG41 and 43.

Transport and Support

Two tactical transport wings — Lufttransport-geschwader 61 at Landsberg and LTG63 at Hohn — operate 76 C160D Transalls, while the Transall OCU at Wunsdorf has a further 14 of the type. A special Government and VIP flight based at Koln-Wahn has four Boeing 707-320Cs, three Lockheed Jetstars and four Hansa Jets. In April 1977 the first of three VFW 614s was delivered to the flight, the second arriving in June and the third aircraft in July. Based at Ahlhorn is the Helicopter Transport Wing HTG 64, flying 105 Bell UH-1Ds. Every Geschwader has assigned to it four Dornier Skyservants for liaison and light transport duties, while for SAR and weapon range liaison duties on Sardinia there are three Bell 212s delivered in 1978.

Nine Pembrokes, three BAC Canberras, two Douglas C-47s and five Nord Noratlas remain in service on communications and special duties. Four HFB320 Hansa Jets have been extensively converted for the ECM role and operate from Lechfeld. The Piaggio P149D still flies with the Air Cadets Regiment in limited numbers for familiarisation flights. A replacement for these aircraft is currently being sought, types under consideration include the Pilatus PC-7 Turbo-Trainer and the RFB Fantrainer. A total of 41 Northrop T-38 Talons and 35 Cessna T-37Bs are stationed in the United States for pilot training.

Title: Marineflieger
Headquarters: Kiel-Holtenau
Strength: 195 aircraft

The Naval Air Arm of the Bundeswehr is scheduled to update its combat force of three fighter-bomber squadrons, currently flying 96 Lockheed F-104G Starfighters, and its reconnaissance unit with 25 RF-104Gs, with 112 Panavia Tornados from 1979. MFG1 at Schleswig and MFG2 at Eggebek are the two Geschwader involved. At Nordholz, MFG3 operates 18 Breguet Atlantics bought in 1966 for MR and ASW duties, and the replacement of these aircraft is currently being considered. For SAR duties, 21 Westland Sea King Mk 41s equip MFG5 at Kiel-Holtenau with detachments at Borkum, Heligoland and Sylt. Twenty Dornier Skyservants fly on communications and liaison with 2/MFG5, together with 15 Sikorsky H-34G helicopters previously flown in the SAR role.

Title: Heeresflieger
Headquarters: Munster
Strength: 550 aircraft

The army air corps provides liaison, communications and observation facilities for the Bundeswehr which is composed of three corps, each having its own Army Aviation Command. Each AAC has a squadron of Alouette IIs, a light transport regiment with two Staffeln of 20 Bell UH-1Ds each and a medium transport regiment with two Staffeln each of 16 VFW-built Sikorsky CH-53Gs. The main headquarters of the Heeresflieger are at Munster in the north, Koblenz in the centre and Ulm in the south. A helicopter training centre at Buckeburg is equipped with all three major rotary-wing types. Dornier-built Bell UH-1Ds number 195, and there are

Top: *German Marine F-104G Starfighters of MFG2 at Eggebek.* / S. G. Richards

Above: *Equipping three German Army transport regiments are VFW-Fokker CH-53Gs.*

240 Aérospatiale Alouette IIs, including about 50 Alouette-Astazous. A total of 110 VFW-Fokker CH-53Gs are in service with the three medium transport regiments, TR15, TR25 and TR35.

A total of 227 MBB Bo105M liaison and communications helicopters are planned to replace the Alouettes, and an initial order for 100 was being met by MBB in 1978. In answer to the Army's PAH-1 anti-tank helicopter requirement orders have been placed for 212 Bo105Ps armed with six HOT anti-tank missiles. Deliveries are due to begin in September 1979, with full entry into service in 1980, when a regiment of 56 PAH-1s will be attached to each German army corps and divided into flights of seven aircraft, which will in turn be attached to brigades or divisions. The army also has a PAH-2 requirement for a specialised anti-tank helicopter and an order has still to be placed.

Ghana

Title: Ghana Air Force
Headquarters: Accra
Strength: 52 aircraft

Established chiefly as a transport and internal policing force, the GAF has, in recent years, undergone a modernisation programme replacing older types. The air force's sole combat aircraft, seven Aermacchi MB326Fs purchased in 1965, are based at Tamali in Northern Ghana and were joined in 1977-8 by a further six single-seat MB326KB aircraft. At Takoradi on the coast are six Short

Above: *Fokker F27 of the Ghana Air Force.* / Fokker-VFW

Skyvan 3Ms used for tactical support, communications, coastal patrol and casualty duties; eight B-N Islanders arrived in 1973-74 for paratroop training, aerial survey and casevac duties. A transport squadron operates three Fokker F27 Mk 400Ms and two F27 M 600s on longer range duties and SAR; a single F28 Mk 3000 is used for VIP flights.

In 1973, the first six Scottish Aviation Bulldog 120 primary trainers were received followed two years later by seven more and these currently operate from Takoradi. A helicopter squadron has four Aérospatiale Alouette IIIs and two Bell 212s, the latter used for VIP and communications.

Great Britain

Title: Royal Air Force
Headquarters: London
Strength: 1,400+ aircraft

Combat

Although Britain's main strategic deterrent is vested in the Royal Navy's Polaris submarines, the Royal Air Force provides an airborne back-up in the form of Strike Command's No 1 (Bomber) Group equipped with six squadrons of Hawker Siddeley Vulcan B2 medium bombers. Totalling 84 aircraft, the six units are Nos 9, 35, 44, 50, 101 and 617 Squadrons plus No 230 OCU home-based at Scampton and Waddington and flying chiefly in the long-range low-level penetration role. Supporting these aircraft and the interceptor units are 24 Handley Page Victor K2 tankers flying with Nos 55 and 57 Squadrons and No 232 OCU at Marham. Four Vulcan SR2s equip No 27 Strategic Reconnaissance Squadron at Scampton, and at Wyton No 39 Squadron operate BAC Canberra PR9s in the high altitude reconnaissance role alongside the electronic surveillance unit, No 51 Squadron, which has three Nimrod R1s and four Canberra B6s. Also at Wyton is the Electronics Warfare School and the joint-Services-operated No 360 Squadron, with 17 ECM Canberra T17s. The Canberra training unit, No 231 OCU, is based at Marham with the special target-facilities unit, No 100 Squadron, also Canberra-equipped with B2s, E15s, T19s and T4s mainly in the high-altitude role; the other target-facilities unit is No 7 Squadron with Canberra B2s, T4s and TT18s at St Mawgan. The Hawker Siddeley Buccaneer S2A/2B wing at Honington has nearly 50 aircraft equipping Nos 12 and 208 Squadrons, the former assigned the maritime strike task; the Buccaneer OCU, No 237, is also on the base. Replacing both the Vulcans and Buccaneers in the 1980s will be 220 Panavia Tornado GR1s of 385 planned for procurement starting in 1979. First Tornado base will be Cottesmore, which will accommodate an international OCU. The remaining 165 aircraft will be Tornado F2s, planned to take over interceptor duties from the Phantoms and Lightnings in the 1980s. To supplement the Victor tankers, five Standard and four Super VC10s are being converted into tanker configuration by British Aerospace. Also operated within No 1 Group is Brize-Norton-based No 115 Squadron flying six HS Andover E Mk 3s on calibration duties.

Air defence is the responsibility of 11 Group, Strike Command, with headquarters at Bentley Priory. The BAC Lightning F3/F6/T5 continues in service with No 5 and 11 Squadron at Binbrook, nearly 40 aircraft making up the force and scheduled to remain in use until 1984. Air defence mainstay is the McDonnell Douglas F-4K/M Phantom which in its FGR2 form equips Nos 23, 29, 56 and 111 Squadron and No 228 OCU at Coningsby, Wattisham and Leuchars, while ex-RN FG1s equip No 43 Squadron at Leuchars. Tied in with the interceptors are a number of missile units of the RAF Regiment including No 27 Squadron with Rapier SAMs for the protection of Leuchars and Lossiemouth, and No 85 Squadron, RAF, with Bloodhound 2s at Bawdsey, North Coates and West Raynham. A further unit in 11 Group, is the early-warning Shackleton AEW2 squadron, No 8, flying from Lossiemouth with 12 aircraft. These old aircraft are due to be replaced by 11 HS Nimrod AEW3s ordered in 1977 and planned for delivery in 1982.

No 18 (Maritime) Group has responsibility for SAR

and MR over the North Sea, Atlantic and home waters. Four squadrons (Nos 42, 120, 201 and 206 plus No 236 OCU) with 35 Nimrod MR1s operate from St Mawgan and Kinloss. The total Nimrod inventory is 46 aircraft and most will undergo a modernisation programme from Mk 1 to Mk 2 beginning in 1979. Helicopter SAR detachments are based around Britain from two squadrons equipped with Westland Whirlwind HAR10s and Wessex HC2s. No 22 Squadron has five flights, one each at Chivenor (A Flight), Leuchars (B Flight), Valley (C Flight), Brawdy (D Flight) and Manston (E Flight); No 202 Squadron operates four flights, one each at Boulmer (A Flight), Leconfield (B Flight), Coltishall (C Flight) and Lossiemouth (D Flight). To replace the Whirlwinds 15 Sea King HAR3s are being delivered between 1978-80 plus some additional Wessex.

No 38 Group forms the RAF's tactical element and provides air support for the Army. Hawker Siddeley Harrier GR3s equip No 1 Squadron and No 233 OCU at Wittering, and a wing of three squadrons of BAC Jaguar GR1 strike aircraft operate from Coltishall, comprising Nos 6, 41 and 54 Squadrons. Of the 89 single- and 20 two-seat Harriers procured by the RAF, around 80 survive to be joined by a further 24 GR3s between 1978 and 1981. The Jaguar force comprises 163 single- and 37 two-seaters, the majority flying with units in RAF Germany (see below). At Odiham the 38 Group tactical helicopter force is made up of No 72 Squadron with 20 Wessex HC2s, and Nos 33 and 230 Squadrons flying some 36 Puma HC1s. Starting in August 1980, this element is expected to absorb 33 Boeing Vertol CH-47D Chinook medium-lift helicopters ordered in 1978.

RAF Germany's combat units are under the control of NATO's 2nd Allied Tactical Air Force and tasked with nuclear and conventional strike, support, reconnaissance and air defence. The Harriers equip Nos 3 and 4 Squadron at Gutersloh while Jaguar strike operations are conducted from Bruggen by Nos 14, 17, 20 and 31 Squadrons; the fifth Jaguar unit, No 2 Squadron flies tactical reconnaissance from the main PR base at Laarbruch. At the same base, Nos 15 and 16 Squadrons operate Buccaneer S2s in the low-level attack role. Interception duties are flown by No 19 and 92 Squadrons from Wildenrath using Phantom FGR2s, while Command liaison duties are performed by 12 BAC Pembroke C1s of No 60 Squadron at the same base. Based at Gutersloh and assigned to support the First British Army Corps in Germany are three flights of Wessex HC2s of No 18 Squadron totalling 15 aircraft. Protecting the German bases are Rapier and Bloodhound units; RAF Regiment Rapier squadrons are No 63 at Gutersloh, No 16 at Wildenrath, No 26 at Laarbruch and No 37 at Bruggen. Three bases are covered by flights of Bloodhounds of No 25 Squadron. In Hong Kong and based at Sek Kong is No 28 Squadron equipped with Wessex HC2 helicopters.

Transport and Support

In the long-range transport role and based at Brize Norton are 11 BAC VC10s of No 10 Squadron, while tactical and strategic duties are performed by a Lockheed C-130K Hercules C1 wing comprising Nos 24, 30, 47 and 70 Squadrons and No 242 OCU flying from Lyneham. The 45 Hercules in service are undergoing a modernisation programme and most will have a fuselage 'stretch' incorporated in them to increase their capacity. Two communications squadrons, Nos 32 and 207, fly Pembrokes, four HS125-400s and two -600s, 13 DH Devons and four Whirlwinds, while the Queen's Flight at Benson has three HS Andover CC2s and two Wessex HCC4s; 12 Andovers remain in RAF service, both C1s and CC2s.

The training side of RAF Support Command has a strength of 180 BAC Jet Provost T3/T5s, 50 HS Gnat T1s, 131 SA Bulldogs, 19 Dominies, 22 Hunters and 10 Gazelle helicopters, although numbers actually in service are somewhat lower due to attrition. No 4 FTS Valley operates Gnats, 20 Hunters and an increasing number of Hawks which are to replace the Gnats; navigation training is performed by No 6 FTS at Finningley on HS Dominies. A multi-engine training unit based at Leeming uses eight SA Jetstreams, and at CFS Shawbury helicopter tuition is flown on the Gazelles. The Central Flying School is now based at Leeming

Below: *RAF Vulcan B2 of No 101 Squadron.* / Crown Copyright

with Jet Provosts while No 3 FTS, also with Jet Provosts, operates from Dishforth. No 2 FTS at Church Fenton conducts primary flying grading on Bulldogs and No 1 FTS at Linton-on-Ouse has JPs for basic training. A Tactical Weapons Unit based at Brawdy has Hawk T1 weapon trainers and 50 Hunters in three 'shadow' squadrons, Nos 63, 79 and 234. A second TWU is temporarily based at Lossiemouth with a further 30 Hunters to relieve congestion at Brawdy as the Hawks arrive.

Title: Fleet Air Arm
Headquarters: London
Strength: 200+ aircraft

Combat

Three anti-submarine cruisers are planned for Royal Navy service from the early 1980s, HMS *Invincible* (19,500 tons) having been launched in May 1977, and presently fitting out, to be followed by HMS *Illustrious* and *Indomitable*. Forming the main equipment on-board these carriers will be British Aerospace Sea Harrier FRS1 strike fighters, 34 being on order for delivery from 1979, each vessel accommodating one squadron plus one squadron of Sea King ASW helicopters. The three Sea Harrier units being Nos 800, 801 and 802 Squadrons, with the main shore-base being RNAS Yeovilton, where 700H Intensive Flying Trials Unit will be established in 1979. HMS *Hermes* has been converted to the ASW role and will operate No 814 Squadron Sea Kings and a few Sea Harriers; she retains a secondary commando role. HMS *Bulwark* is being restored to full operational status as an ASW carrier while the Navy's only strike carrier, HMS *Ark Royal* was paid off at the end of 1978. Prior to retirement, *Ark's* Air Group comprised No 892 Squadron with Phantom FG1s, No 809 Squadron with Buccaneer S2s, No 824 Squadron with Sea King HAS1s, and B Flight of No 849 Squadron with Gannet AEW3s. The Phantoms and Buccaneers are being absorbed into the RAF while the Gannets will be retired from use.

A Wessex HAS3 anti-submarine helicopter drawn from No 737 Squadron is assigned to each of the six County-class guided missile destroyers and the cruiser HMS *Blake* operates a Sea King of No 820 Squadron. The total Sea King inventory stands at 56 Mk 1s and 21 Mk 2s. One Wasp HAS1 from No 829 Squadron is assigned to each of the Leander, Amazon, Sheffield, Tribal and Rothesay class ships, making 40 flights in all. Replacing the Wasps are Westland Lynx HAS2s, of which there are 88 on order, the first squadron being 702 which supplied the first Lynx Flight to HMS *Birmingham* in 1978. Home base for the squadron and shore HQ of the FAA is RNAS Yeovilton which also houses two squadrons and a training unit of Commando Wessex HU5s (Nos 845, 846 and 707 Squadrons). Prestwick in Scotland forms the base for No 819 Squadron flying Sea Kings. For Royal Marine use, 15 Westland HU4 Commandos were ordered in 1978 and these will join Nos 845, 846 and 707 Squadrons in 1979.

Above: *Royal Navy Westland Wessex 1 of No 771 Squadron.* / Crown Copyright

Transport and Support

In addition to the above units, RNAS Yeovilton is the base for the Fleet Requirements and Air Direction Unit operated by Airwork Services Ltd, which has Canberra T22s and TT18s and Hunter GA11s. Other shore bases include Culdrose, where No 750 Squadron flies eight Sea Prince T1s for observer training, but shortly to receive 16 BA Jetstream T2s converted from ex-RAF machines and due to replace the Sea Princes. At the same base is No 705 Squadron, with Gazelle HT2 pilot training helicopters; No 706 Squadron, with 10 Sea Kings, and No 771 Squadron SAR School, with Wessex 1s. At Lee-on-Solent, No 781 Squadron, the Navy's communications unit, flies four Sea Herons and eight Sea Devons. Portland acts as a training and shore base for Nos 703 and 829 Squadrons, with Wasps; No 737 Squadron with Wessex HAS3s; and No 772 Squadron fleet requirements unit with Wessex HU5s. An air experience flight of Chipmunks is based at Roborough.

Title: Army Air Corps
Headquarters: London
Strength: 330+ aircraft

Above: *Westland Gazelle AOP helicopter of the Army Air Corps.* / Crown Copyright

Army Aviation is a corps in its own right within the British Army and is divided into regiments, each one comprising two squadrons in each Army division, and each squadron is capable of operating as a number of self-supporting flights. Gazelle deliveries continue, with more than 130 of the 158 on order already in service. About half the Gazelles will be armed for the anti-tank role, the remainder being assigned to liaison and communication. Each of the four army divisions in Germany will receive one Gazelle LOH squadron and one TOW missile-armed Westland Lynx anti-tank squadron. A total of 100 Lynx AH1s are on order to replace the AAC's 120 Westland Scouts, and all those deployed will be equipped with TOW missiles for the anti-tank role. There are five regiments based in Germany, made up of Nos 1, 2, 3, 4 and 9 under the control of No 1 Wing at Detmold. These regiments consist of the following squadrons: Nos 651, 661, 652, 662, 653, 663, 654, 664, 659 and 669. No 2 Wing, AAC, UK Land Forces, based at Wilton, controls 5, 6, 7 and 8 Field Force embodying Nos 655, 656, 657 and 658 Squadrons, together with No 7 Regiment, No 8 Flight at Netheravon. Gazelles are flown in Hong Kong by No 11 Flight which also has a detachment based in Brunei, while in Cyprus, 16 Flight operates the eight remaining Aérospatiale Alouette IIs. About 7 DHC Beavers and 24 Chipmunks continue in Army service for a variety of tasks including transport and training.

Greece

Title: Elliniki Aeroporia
Headquarters: Athens
Strength: 530+ aircraft

Combat

Although she has withdrawn from NATO and the 6th Allied Tactical Air Force since the Turkish invasion of Cyprus in 1974, Greece has indicated that in an emergency she would support the Alliance. The United States supplies military aid to the country in return for the use of some Greek military installations. The Hellenic Air Force is organised into three main commands: Tactical Air, Training and Air

Below: *One of 40 Dassault Mirage F1s of the Royal Hellenic Air Force.* / J. M. C. Guhl

Material. Within the 28th Tactical Air Command are six combat wings (Pterighe), each with up to three squadrons (Mire). No 110 Wing at Larissa flies strike/reconnaissance duties with No 345 Squadron operating 18 LTV A-7H Corsair IIs, and Nos 348 and 349 Squadrons each with 14 Northrop RF-5As. At Nea Ankhialos, is No 111 Wing assigned the day interceptor role with Nos 337, 341 and 343 Squadrons, each equipped with 15 Northrop F-5A/Bs. No 114 Wing is based at Tanagra and embraces two interceptor squadrons, Nos 336 and 342, flying 40 Dassault Mirage FICGs, delivery of which was completed in 1977. A further strike wing is No 115 at Souda Bay, comprising two units with A-7H Corsairs, No 338 and 340 Squadrons while a fighter-bomber wing, No 116, is based at Araxos with nearly 30 Lockheed F-104G/TF-104Gs forming a single squadron, No 335. At Andravidha, No 117 Wing operates 18 McDonnell Douglas F-4E Phantoms in No 339 Squadron and a second unit with a further 20 F-4Es. Eight RF-4E reconnaissance aircraft equip a tactical-recce unit and 18 more F-4Es are being received during 1978 for a possible third strike squadron. A total of 64 Phantoms are in HAF service together with 60 Corsair IIs. In addition five two-seat TA-7H trainers plus a sixth conversion from a single-seater have been ordered for delivery in 1980. Also within TAC is No 363 Squadron at Eleusis equipped with eight ex-Norwegian Grumman HU-16B Albatross amphibians and flown under naval control for MR, ASW and SAR duties.

Transport and Support

Air Material Command operates the transport side of the HAF and controls two main units, Nos 355 and 356 Squadrons, based at Eleusis and flying three types of aircraft. The heavy-lift and troop-transport fleet has 12 Lockheed C-130H Hercules, the last four having been received in 1977; nearly 40 ex-Luftwaffe Nord Noratlas reinforced in 1976 by a further 20 ex-Israeli aircraft; and 30 Douglas C-47s. A single Grumman Gulfstream I is used for VIP flights. Also based at the same airfield are three helicopter units: Nos 357 Squadron with 10 Bell 47Gs; No 359 Squadron with 12 Sikorsky H-19Ds and No 362 Squadron with 14 Augusta-Bell AB205s. Additionally, a further 35 utility helicopters, Bell UH-1s, were delivered to Greece from the USA in 1977 and these may have replaced the H-19s. The USA is also to supply a small number of Boeing Vertol CH-47Cs in the near future.

Training Command has a National Air Academy at Dhekelia where students undergo initial training on 20 Cessna T-41As. For basic jet instruction, No 361 Squadron at Eleusis has 18 Cessna T-37Bs with advanced tuition flown on 40 Rockwell T-2E Buckeyes operated by No 360 Squadron at the same base and delivered in 1976-77. The T-2Es have replaced about 50 Lockheed T-33As used for some years and acquired from ex-German stocks. A number of these redundant aircraft have been converted by Dornier into target tugs. Operational conversion is flown on nine Northrop F-5Bs and four TF-104Gs. Also in service are six Canadair CL-215 amphibious water bombers.

The Greek Navy has a helicopter support force of four Alouette IIIs and a recently formed ASW element equipped with Augusta-Bell AB212ASWs. A total of 12-16 machines have been ordered and delivery was underway in 1978.

The Greek Army has two Rockwell Commander 680s, 15 Piper L-21s and 25 Cessna U-17s. Helicopters include five Bell 47Gs, 40 AB204/205s, and 10 Bell UH-1Ds.

Guatemala

Title: Fuerza Aerea Guatemalteca
Headquarters: Guatemala City
Strength: 82+ aircraft

Administered by the Army, Guatemala's air force has only a nominal combat element, a deficiency which tempered this Central American state's confrontation with neighbouring Belize in 1977. As a result, the FAG is believed to have requested the United States for permission to order a small batch of Northrop F-5Es in 1978. The air force currently operates 13 Cessna A-37B jet counter-insurgency aircraft delivered in batches in 1971, 1974 and 1975. The other jet equipment comprises three Cessna T-37Cs for conversion training, and five Lockheed T-33As for advanced work; both types being supplied by the USA under a military aid agreement.

The Escuadron de Transporte has its headquarters at Guatemala City Airport and operates nine Douglas

Below: *Guatemalan Air Force Arava.* / H. Levy

C-47s, 10 IAI Arava light transports bought from Israel, one C-54 and one DC-6. For liaison purposes there are six Cessna 170s, three Cessna 180s, two Cessna U206Cs, six Bell UH-1Ds, three Sikorsky UH-19s and a Hiller OH-23G. For primary training 10 Aerotec T-23 Uirapurus have been bought from Brazil and there are seven North American T-6s for basic training; at least two Cessna 172s are in use.

Guinea Republic

Title: Force Aérienne de Guinea
Headquarters: Conakry
Strength: 30+ aircraft

This West African republic, for many years under Soviet influence, is moving steadily into a more non-aligned position. However, the air force continues to operate eight MiG-17F fighter-bombers supplied by Russia some years ago and at the time of writing there appears to be no move to replace them with Western types. For transport work there are four Il-14s, four An-14s, two Il-18s for VIP duties, and a single Bell 47G helicopter. Training is conducted on seven Yak-18s, a couple of MiG-15UTIs and some Czech L-29 Delfins.

Guyana

Title: Guyana Defence Force Air Command
Headquarters: Georgetown
Strength: 17 aircraft

Formerly British Guiana, this independent Commonwealth country has an airborne policing force equipped with both fixed-wing and helicopter types. Largest aircraft in use is the Britten-Norman Islander, of which there are eight in service for transport and supply duties, supplemented by a Beech King Air 200 purchased in 1975 and a Cessna U206F delivered the following year for personnel transport. Helicopters total two Aérospatiale Alouette IIIs, three Bell 212s and two Bell 206s.

Haiti Republic

Title: Haiti Air Corps
Headquarters: Port-au-Prince
Strength: 33 aircraft

This small air arm has received limited American aid in recent years, but no modern combat aircraft are in service. The sole fighter-bomber squadron has relinquished its long-serving North American F-51D Mustangs and now flies in the counter-insurgency role equipped with eight Cessna 0-2A Super Skymasters fitted with underwing hard-points for external stores. Transport and support duties are performed by three Douglas C-47s, two Beech C-45s, a single Cessna 402 and three DHC Beavers. The helicopter element has four Sikorsky H-34s and five S-55s for a variety of tasks while training is performed on three T-6s, three Cessna 150s and a Cessna 172.

Below: *Cessna 0-2A of the Haitian Air Corps.*
/ H. Levy

Haute-Volta

Title: Force Aérienne de Haute-Volta
Headquarters: Ougadougou
Strength: 10 aircraft

Receiving its independence from France in 1960, the Upper Volta Republic has a modest air component operating as part of the small 3,000-man army. Almost all the aircraft in use have been supplied from ex-French military stocks and no pure combat types have yet been purchased, the force flying transport, supply and liaison duties. Two Douglas C-47s were supplemented in September 1977 by a single Hawker Siddeley HS748, joining three MH1521M Broussards, an Aero Commander 500 and a Reims C337 Super Skymaster. Two Nord 262s are flown on VIP and utility duties.

Honduras

Title: Fuerza Aerea Hondureña
Headquarters: Tegucigalpa
Strength: 52 aircraft

The largest of the Central American nations, Honduras has had frequent confrontations with neighbouring El Salvador and to match more modern jet equipment acquired by the latter country, the Honduran Air Force purchased 12 Dassault Super Mystere B2 fighter-bombers from Israel in 1976. These ex-IAF aircraft, now based at Tocontin, have been re-engined with American Pratt & Whitney J52s in place of their original French Snecma Atars and have significantly increased Honduran air power. The FAH is also equipped with a squadron of six Cessna A-37Bs, delivery of which was made in three batches between June and November1975. There is also a reconnaissance/trainer unit based at San Pedro Sula airbase equipped with three Lockheed RT-33As.

Modern equipment has also been received for the transport force in the form of two IAI Arava utility aircraft delivered in 1976. They fly alongside at least five Douglas C-47s, two C-54s and a number of Beech C-45s. Four Cessna 180/185s fly liaison duties, while a basic training unit is equipped with five Cessna T-41As delivered in 1973, about ten T-28 Trojans and six North American T-6s. Only three Sikorsky H-19s are flown.

Below: *Six Cessna A-37Bs were delivered to the Honduran Air Force in 1975.* / Cessna

Hong Kong

Title: Royal Hong Kong Auxiliary Air force
Headquarters: Kai Tak
Strength: 7 aircraft

This policing and communications force, a department of the HK Government, is based near RAF Kai Tak alongside the international airport. It is

Below: *Hong Kong Government B-N Islander.*

manned chiefly by part-time volunteers but operates full-time. Three Aérospatiale Alouette III helicopters fly SAR, liaison and internal security patrols, and a single Britten-Norman Islander, delivered in 1972, is used for anti-smuggling and survey work. Two Scottish Aviation Bulldog 128 trainers were delivered in 1977 and provisional plans call for the phase-out of the single Beech Musketeer in 1980.

Hungary

Title: Hungarian Air Force
Headquarters: Budapest
Strength: 200+ aircraft

Combat

The smallest of the Warsaw Pact air arms, the HAF is closely supervised by the Soviet Air Army, elements of which are based in the country. In the interceptor role, two fighter regiments with three squadrons each, have about 100 MiG-21SMTs including perhaps a dozen two-seat MiG-21U conversion trainers. For pure ground-attack work there is a fighter-bomber regiment of three squadrons flying 36 Su-7MBs.

Transport and Support

Approximately 30 transport aircraft are in service, the majority being An-24s and An-26s having replaced most of the An-2s and Il-14s; two Tu-134s operate in the VIP/government role. The helicopter elements have about 15 Mi-4s and Mi-8s, together with a small number of Ka-26 co-axial utility machines. Training at the primary stage is flown on Zlin 42s, students progressing to Yak-18s, Czech L-29 Delfins and MiG-15UTI advanced trainers.

India

Title: Indian Air Force
Headquarters: New Delhi
Strength: 1,500+ aircraft

Combat

India's combat elements, which have some 900 aircraft in 36 squadrons, operate a mixture of Soviet and Western types, many of which are approaching obsolescence requiring the Government to seek urgent replacements. Backbone of the interceptor force is the licence-built MiG-21 produced by Hindustan Aircraft. Fifty Soviet-supplied FL examples were followed by 196 built by HAL and supplanted on the production line by the more up-to-date MiG-21MF, 150 being built for the 11 squadrons currently flying MiGs (Nos 1, 3, 4, 8, 28, 29, 30, 45, 47, 108 Squadrons and an OCU flying MiG-21U two-seat trainers). From 1980-81 India will build the advanced MiG-21bis under licence. Supplementing the MiGs are some 150 HAL-built Gnat F1 light fighters equipping five squadrons (Nos 2, 15, 18, 22 and 23) to be joined over the next few years by up to 100 of the new, improved performance Mk 2 Gnat, named Ajeet. The first unit, No 9 Squadron, with the new type is due to become operational in 1979; a two-seat trainer Ajeet is under development for service entry in 1981.

Four ground-attack squadrons have about 130 Hawker Hunter F56/T66s and three further units have some 100 indigenous Hindustan HF-24 Marut Mk 1 strike aircraft, plus a batch of ten two-seat Mk 1T conversion trainers. Of 150 Sukhoi Su-7MB fighter-bombers delivered to India by the Soviet Union, about 120 remain in service with four squadrons (Nos 32, 101, 221 and 222), while

Below: *MiG-21 interceptor of the Indian Air Force.*

Above: *Indian Navy Westland Sea King Mk 42, one of 15 in service.* / Westland

86 BAC Canberra B(I)58s, B74s, B(I)12s and T13s fly with three bomber units, Nos 5, 16 and 35 Squadrons. A single PR unit, No 106 Squadron, has 12 Canberra PR7s. To replace the Su-7s, Canberras and Hunters, the IAF issued a requirement for a Deep Penetration Strike Aircraft which has been met by the BAC/Breguet Jaguar International. Reports have stated that up to 60 aircraft would be bought from the UK production line followed by the licence-production of a further 140 in India.

Transport and Support
For heavy logistic work, particularly in the northern Himalayan areas, there are two squadrons with 30 Russian An-12 freighters (Nos 25 and 44) supplemented by three medium transport squadrons flying 40 Fairchild C-119G jet-boosted Packets (Nos 12, 19 and 48); 40 Douglas C-47s survive in Nos 11, 43 and 49 Squadrons. STOL operations are flown by No 33 Squadron with 20 DHC Caribou and by a few remaining DHC Otters of No 41 and 59 Squadron. By far the largest type numerically in IAF transport service is the HAL-built HS748M with a total of 81 in use or ordered, the majority flying in No 12 Squadron. Six more operate with a Headquarters/VIP flight alongside one Tupolev TU-124.

About 200 helicopters perform a variety of tasks including transport, SAR, supply and liaison, with three squadrons equipped with 50 Mil Mi-4s, three squadrons with 35 Mi-8s, 120 Alouette IIIs in three units and 12-15 Bell 47Gs, the latter recently supplanted by Alouette IIIs in the training role. Deliveries are under way of 100 licence-built Aérospatiale SA315 Cheetahs to replace the 40-odd home-designed Krishak fixed-wing AOPs used by the Army. A few Alouette IIIs and about 20 Auster AOP9s also fly AOP duties.

IAF Training Command, with HQ at Bangalore, has about 70 HT-2 primary trainers (scheduled to be replaced by the new Hindustan HPT-32 in 1981-82) based at Bidar, and about half the 130 HJT-16 Kirans on order, flying at the Air Force Academy at Dundigal. A Fighter Training Wing at Hakimpet operates 50 Polish TS-11 Iskras, delivered in 1975-76 to supplement the slow delivery of Kirans, with students flying conversion on two-seat Hunter T66s, Canberra T13s, MiG-21Us and Su-7Us at operational level. Multi-engine training is flown on HS748s at Yelahanka, near Bangalore, while navigation training is flown on five further HS748s plus two signals training versions.

Title: Indian Navy
Headquarters: New Delhi
Strength: 125 aircraft

Combat
The Navy operates the aircraft carrier INS *Vikrant* (ex-RN HMS *Hercules*, bought in 1957), which can accommodate 18 of No 300 Squadron's 25 Hawker Sea Hawk fighter-bombers. A replacement for these ageing aircraft is long overdue and an order for six single-seat Sea Harriers and two trainer versions is thought likely. An eventual total of 25 Sea Harriers is envisaged. *Vikrant* carries in addition to the Sea Hawks, four Breguet Alizé ASW aircraft and two Alouette III planeguard helicopters from a total of 20 Alizés in No 310 Squadron and 10 Alouettes in

No 321 Squadron. Other helicopter units include No 331 Squadron flying eight torpedo-armed Alouette IIIs for deployment aboard Leander-class frigates, and Nos 330 and 336 Squadron based at Cochin Naval Base equipped with 15 Westland Sea King Mk 42 ASW machines.

Based at Goa is No 312 Squadron flying five Lockheed Super Constellations on maritime patrol duties and due to be withdrawn in 1981, while No 315 Squadron at Dabolim operates three Ilyushin Il-38 ASW aircraft delivered in 1977 with a possible re-order for another three in the near future. A further acquisition from Russia is a batch of five Kamov Ka-25 helicopters which are believed to be destined for the two Kashin-class destroyers due for delivery in 1978 and will be operated by No 331 Squadron.

Transport and Support

For fixed-wing training, No 551 Squadron has a complement of seven HJT-16 Kirans of 15 ordered, four DH Vampire T55s and four Sea Hawks. At Cochin NAB, No 550 Squadron has two DH Devons for liaison flights and five Britten-Norman Defenders for training and communications, the latter aircraft were delivered in 1977 and are being modified for coastal patrol duties with the addition of nose radar. Four Hughes 300s and some Alouettes are used for helicopter training and form No 561 Squadron.

Indonesia

Title: Tentara Nasional Indonesia Angkatan Udara
Headquarters: Jakarta
Strength: 173+ aircraft

Combat

During the 1960s, the Indonesian Air Force was regarded as one of the most powerful in south-east Asia, having been supplied with aircraft and equipment by the Soviet Union. However, a change of government and policy resulted in the severing of links with Russia with the consequent withdrawal from use of all the Soviet aircraft. Military aid has been forthcoming from the USA and particularly Australia who supplied a squadron of 16 ex-RAAF Commonwealth Sabre fighters in 1972. These are presently based in East Java and form the TNIAU's main interceptor force. 12 Northrop F-5Es and four F-5Fs are on order to replace these aircraft. A second squadron is equipped with 16 Rockwell OV-10F Bronco counter-insurgency aircraft delivered in 1976-77, while a third unit still operates 14 North American F-51D Mustangs in the ground-attack role. Soviet types grounded and in storage total 22 Tupolev Tu-16 bombers, ten Ilyushin Il-28 light bombers, 35 MiG-19s, 15 MiG-21s, 40 MiG-15/17s, 10 Il-14 transports, 10 An-12s, 20 Mi-4s and nine Mi-6s.

Transport and Support

Eight Lockheed C-130B Hercules provide heavy-lift capacity and operate with No 31 Squadron while No 2 Squadron has 12 Douglas C-47s and three Short Skyvans. Eight Fokker-VFW F27 Mk 400Ms were ordered in 1975 to modernise the force and six GAF Nomad Mission Masters have been acquired through the Indonesian/Australian aid agreement. Nurtanio, the country's aircraft industrial concern, is delivering 25 Casa C212 Aviocar light transports to the Service and is also to assemble Aérospatiale SA330 Puma helicopters under licence, six aircraft being on order for the TNIAU. A Lockheed JetStar 6 is used for VIP flights, and a variety of smaller types are flown on liaison and communications work, including seven DHC Otters, five Cessna 401s, two Cessna 402s and five Cessna T207s. The helicopter fleet includes four Alouette IIIs, two Bell 204Bs, four Sikorsky H-34Ds and an S-61A; a number of ex-Australian Army Bell 47Gs were received during 1978, and some Bell 205s are expected to be ordered.

Modernising the training elements has been of considerable importance in the TNIAU. Ordered via Hawker-de Havilland in Australia and delivered in 1978 were 16 Beech T-34C Turbo-Mentor basic trainers, replacing 15 older T-34A Mentors, while British Aerospace received an order for eight Hawk Mk 53 advanced trainers in April 1978. The Hawks will replace 10 Lockheed T-33As presently in service and a repeat order is likely to follow.

Below: *An L-29 Delfin of the Training Academy, Indonesian Air Force.*

Title: Angatan Laut Republik Indonesia
Headquarters: Jakarta
Strength: 33 aircraft

Six GAF Nomad Search Master B utility aircraft are in service with the naval air arm, having been acquired in 1975 under the aid agreement with Australia. They are used for coastal patrol and transport work from Sourabaya and are being joined by a further six aircraft for a second unit. Five Grumman HU-16B Albatross amphibians are still operated on SAR duties, while six Douglas C-47s fly in the support and transport roles. Helicopters used for liaison work include three Alouette IIs, three Alouette IIIs and four Bell 47Gs.

Title: Tentara Nasional Indonesia Angatan Dorat
Headquarters: Jakata
Strength: 30+ aircraft

The air arm of the Indonesian Army was formed in 1959 for liaison and support duties. Sixteen Bell 205A-1s were delivered from the USA in 1977 for utility duties and complement a variety of fixed-wing and helicopter types: two Douglas C-47s, two Aero Commander 560s, one DHC Beaver and a Beech 18, seven Alouette IIIs and some Cessna 185s, 0-1s, 310Ps and a number of licence-built Polish Wilga 32s.

Iran

Title: Iranian Air Force
Headquarters: Tehran
Strength: 843 aircraft

Combat

This once modest air arm now stands as one of the world's largest air forces having undergone a massive modernisation programme since the early 1970s. However, the new Islamic government has halted the arms build up and cancelled many large orders placed with Britain and the USA. The lack of American assistance in the operation of some of the IAF's more sophisticated equipment will doubtless reduce their efficiency. The most sophisticated aircraft in use with the IAF's combat units is the Grumman F-14A Tomcat variable-geometry interceptor. Delivery of 80 Tomcats was completed in 1978 to two bases, Khatami AFB at Isfahan where 50 are based, and Shiraz farther south, which accommodates 30 aircraft. Other elements of the current IAF strike-interceptor force comprise 10 squadrons of Phantoms based at Shiraz, Tabriz and Tehran comprising 32 F-4Ds, 141 F-4Es and four RF-4Es, plus a further 12 RF-4Es and 36 F-4Es delivered between May 1976 and July 1977. Eight fighter-bomber squadrons operate 141 Northrop F-5Es and 28 two-seat F-5Fs from Buskehr on the coast and from Tabriz, having replaced 117 older F-5A/RF-5As and 22 F-5Bs which have been passed on to other air forces. For maritime reconnaissance along Iran's coastline there are six Lockheed P-3F Orions based at Bandar Abbas.

Transport and Support

To support the large combat element, the IIAF operates 13 Boeing 707-320C tanker/transports, six equipped for three-point refuelling, and three Boeing 747s similarly equipped. A further three 747s are straight transports and the force has placed an order for four more 747F versions for freight duties. The tactical element has 15 Lockheed C-130E Hercules and 49 C-130H Hercules with four aircraft converted into signal monitoring machines work along the border with Russia. From Holland, Iran has purchased 13 Fokker-VFW F27 Mk 400M Troopships, including two for aerial survey, and five F27 Mk 600s, three of the latter for VIP use.

Iran's helicopter force by the mid-1980s would have totalled more than 1,100 machines, most operated by the Army and almost all bought from America and Italy. However, the political crisis has forced a number of cuts in orders and the present strength is as follows:

For SAR duties with the Air Force there are 39 Bell 214C utility machines delivered in 1976-78, two Agusta-built Sikorsky S-61A-4 VIP mahines, six Bell

Below: *In worldwide use, the Northrop F-5E. This example is in Imperial Iranian Air Force insignia.*

214As for government use, two Meridionali CH-47C Chinooks for heavy freight work, 10 Kaman HH-43F Huskies, 16 Aérospatiale Super Frelons, 45 Agusta-Bell AB205s, five AB212s and 70 AB206 liaison helicopters.

Fixed-wing liaison flights are flown by three Aero Commander 690s, four Dassault Falcon 20s and seven DHC Beavers. A total of 49 Beech Bonanzas — 39 F33Cs and 10 F33As — are in service for training.

Title: Iranian Navy
Headquarters: Tehran
Strength: 61 aircraft

Like the other two services, the Navy air arm was expanded considerably under the Shah's rule and now operates 10 Agusta SH-3D Sea King anti-submarine helicopters with a further ten on order. Six Sikorsky RH-43D mine-sweeping helicopters are being delivered during 1978 and will join five AB205s, 14 AB206s and six AB212s. Six Rockwell Shrike Commanders operate in the liaison role, and two Fokker-VFW F27 Mk 400Ms and two Mk 600s fly transport work based at Mehrabad.

Title: Iranian Army
Headquarters: Tehran
Strength: 537 aircraft

The largest type in service with the aviation element of the Iranian Army is the Boeing Vertol CH-47 Chinook. A total of 20 have been received from the parent company and a further 70 from the Italian licensee, Meridionali. Most are used for tactical assault duties and heavy freight work. For anti-tank use, there are 202 Bell AH-1J Cobras, many fitted with TOW wire-guided missiles, while for liaison and general duties, 287 Bell 214As were delivered to the service, order completion occurring early in 1978.

The Army also has a fixed-wing element operating about 40 Cessna 185s, 10 Cessna 0-2s, six Cessna 310s and two Fokker-VFW F27 Friendships — one Mk 400M and one Mk 600 for transport and target-towing duties.

Below: *A total of 20 Agusta-built AS-61ASW helicopters are being delivered to the Iranian Navy.* / Agusta

Bottom: *Cessna 0-2 of the Iranian Army.* / Cessna

Iraq

Title: Iraqi Air Force
Headquarters: Baghdad
Strength: 530+ aircraft

Combat

For the past few years, Iraq has relied strongly on military aid from the Soviet Union to equip her armed forces. However, this reliance is becoming less critical with the country's gradual thaw with western nations, notably France with whom she has placed large orders for aircraft and equipment, and her future Air Force inventory will show a marked spread of diverse types from both east and west. The most important purchase from France is a batch of Dassault Mirage F1E air superiority fighters, comprising 32 F1EQ single-seaters and four F1BQ two-seat combat trainers, ordered in 1977 and planned for delivery in 1979-80. These are expected to join Air Defence Command which has a ground-based early-warning system integrated with five squadrons of 90 MiG-21PFMs, of which two units have the -21MF versions.

The IAF's Support Command operates two strike squadrons with 40 MiG-23/27 Flogger variable-geometry aircraft, the -23 predominating; three fighter-bomber squadrons with 30-plus MiG-17s; and a long-range bomber squadron with 12 Tupolev Tu-22s delivered in 1973. The following year the IAF received two squadrons of Sukhoi Su-22 v-g strike aircraft. More than 30 HS Hunter FGA9/FR10s equip three further units of 46 Hunters originally delivered, and remaining elements in Support Command total three Su-7B strike units with 50 aircraft and a light bomber squadron with 10 Il-28s of doubtful operational status. For short-range counter-insurgency duties there are 16 BAC Jet Provost T52s out of 20 delivered in 1964-65.

Transport and Support

Fixed-wing transport elements in the IAF are mostly equipped with Soviet aircraft, the most modern type being the new Il-76 STOL aircraft, delivery of which was expected in 1978 although exact numbers are not yet known. Six An-12 freighters form the heavy-lift component supported by 13 Il-14s, 10 An-24s, at least two An-26s, 12 An-2s, two DH Herons, two Tu-124s for VIP use and four Britten-Norman Islanders for light utility work; a Dassault Falcon 20 is also used for VIP flights. The helicopter force has Russian, French and British types and totals nearly 180 machines. Delivered in 1977 were 10 Aérospatiale Super Frelons (reduced to nine following the loss of one during training) and 40 SA342L Gazelle light helicopters, the latter used for anti-tank duties and fitted with SS11s, SS12s and HOT missiles. The rest of the fleet totals four Mi-1s 35 Mi-4s, 36 Mi-8s (two in VIP layout), a few Mi-6s, 40 Alouette IIIs, two VIP SA330 Pumas and 12 Westland Wessex 52s.

Trainers include Yak-11s, MiG-15UTIs, Czech L-29s — gradually being replaced in the advanced trainer role at the Air Force College at Tikret by 24 improved L-39s — MiG-21Us and Hunter T69s. Announced in mid-1978 was an order for 48 AS202/18A Bravo ab initio trainers for the Air Force and Iraq Flying Association.

Below: *Iraqi Air Force L-39 Albatross advanced trainer.*

Ireland (Eire)

Title: Irish Army Air Corps
Headquarters: Baldonnel
Strength: 41 aircraft

The increased pressure of border operations and the advancing years of the equipment flown by the IAAC have prompted a re-equipment programme which is

Beech Super King Air 200 of the Argentine Navy.
/ R. Alexander

Right: *Embraer R-95 Bandeirante of the Forca Aerea Brasileira.* / Embraer

Below: *General Dynamics F-111Cs of the Royal Australian Air Force.* / Martin Horseman

Below right: *Cessna A-37Bs of the Fuerza Aerea de Chile.* / Cessna

41

J-630
J-629
629

Above: *Northrop F-5A of the Blue Diamonds aerobatic team, Philippine Air Force.* / A. Anido

Below: *McDonnell Douglas F-15 Eagles of the United States Air Force.* / Martin Horseman

now nearing completion. To replace some long-serving DH Vampire T55s, six Aérospatiale Super Magisters were delivered to the Service during 1976. Although armed with 7.62mm machine-guns and underwing rockets these aircraft are primarily used for training and are the only jets in the Air Corps. Based at Baldonnel with the Magisters are 10 Siai-Marchetti SF260W Warriors, delivered in 1977 and replacing Chipmunks in the basic training role. The six surviving Chipmunks will remain in use for liaison staff flying as long as spares last. Three DH Doves are still flying, one equipped with cameras for aerial survey work with the Photographic Unit, and the IAAC has to make a decision on a replacement for these aircraft soon. One possibility is the acquisition of a Beech Super King Air 200, one of which was leased during 1977-78 for coastal fisheries protection.

Above: *Irish Air Corps Rheims Rocket with underwing rocket pods.*

The helicopter element comprises eight Alouette IIIs of the Helicopter Rescue Service, the purchase of these machines beginning in 1963. Purchased in 1972 were eight Cessna-Reims FR172Hs for the Army co-operations unit which patrols the borders, flying from the base at Gormanston, County Meath.

Israel

Title: Israel Defence Force/Air Force
Headquarters: Tel Aviv
Strength: 1,072+ aircraft

Combat

American arms supplies to Israel continue with present orders extending equipment deliveries to 1983. This time-scale allows for the delivery of 30 more McDonnell Douglas F-15 Eagles to join the 25 already received or on order. Also approved by the US Government is Israel's purchase of 75 General Dynamics F-16A/B interceptors to help balance arms supplies by the USA to Egypt and Saudi Arabia. The single IDF/AF Eagle squadron is operational and includes four refurbished development aircraft delivered in December 1976, although the remaining machines are new.

The McDonnell Douglas Phantom force, in service since 1969, totals officially 204 F-4E fighter-bombers and 12 RF-4E reconnaissance aircraft in some seven strike/interceptor squadrons. Most of the F-4Es have been fitted with leading-edge slots to improve manoeuvrability. More than 250 McDonnell Douglas A-4E/F/H/N Skyhawks, plus some 24 two-

Below: *McDonnell Douglas F-15A Eagles of the Israeli Air Force.* / Israir

seat TA-4E/Hs, operate in the ground-attack role with six squadrons, supplemented by an increasing number of Israeli Aircraft Industries Kfir C-2 multi-role jets. A derivative of the Dassault Mirage III but powered by an American J79 engine, the Kfir was revealed in 1975 and an IDF order for 100 aircraft is currently being met. A few interim-standard SNECMA Atar 9C-powered aircraft known as Neshers fought in the 1973 war and remain in use. Of the 72 Dassault Mirage IIICJ interceptors and five two-seat IIIBJs originally supplied by France before the 1967 war, about 40 aircraft survive in two squadrons.

For electronic intelligence and battlefield reconnaissance duties, the IDF has two Grumman EV-1 Mohawks, equipped with SLAR pods and delivered in 1976, and four Grumman E-2C Hawkeye electronic surveillance and early warning aircraft, delivery of these machines being completed in mid-1978.

A batch of 30 Hughes 500M-D Defender armed helicopters was ordered in 1978 for the anti-tank role armed with TOW missiles. The order was expected to go to the Bell Cobra but the Hughes machine was selected for cost reasons; delivery of these aircraft is planned for mid-1979.

Transport and Support

Tactical transport squadrons have a fleet of Lockheed Hercules comprising 12 C-130Es, 12 C-130H and two KC-130H tankers; 10 Boeing 707s converted by IAI into tanker/transports and command/control aircraft; 18 Douglas C-47s (some used as navigation trainers); and more than six Boeing C-97s flying as three-point tankers, ECM aircraft and freighters.

A large fleet of helicopters includes eight Aérospatiale Super Frelons re-engined with American powerplants in place of their Turbomeca turboshafts, 29 Sikorsky S-65/CH-53Ds for assualt and transport, 45 Bell 205s, 20 Bell 206 JetRangers for training and liaison and 12 Bell 212s.

For short-range utility and liaison work the force has 14 Dornier Do28s and 35 Do27 observation aircraft supplied by West Germany as war reparations, 28 Cessna U206C Super Skywagons, 16 Beech Queen Air 80s acquired in 1974, 8-10 B-N Islanders and one IAI Westwind 1123 for VIP duties. Contrary to reports, there are no Aravas currently in IDF service although three were used during the 1973 war for casevac duties.

Training is the task of some 80 IAI-built Fouga Magisters which also double in the light strike role, primary grading being flown on 20 Piper Super Cubs, two Cessna T-41Ds and a single Cessna 180. The Israeli Navy operates three 1124N Sea Scan versions of the Westwind for MR and patrol duties, the aircraft being flown by Air Force crews.

Italy

Title: Aeronautica Militare Italiano
Headquarters: Rome
Strength: 1,000+ aircraft

Combat

One of the original signatories of the North Atlantic Treaty, Italy plays an important role in Allied Forces Southern Europe, having some 19 squadrons assigned to 5th ATAF while retaining a small number of units for national defence. The current AMI force allocated to NATO is made up chiefly of Lockheed Starfighters, the Italian-built F-104S version now predominating and having almost replaced the older F-104G. A total of 205 Aeritalia-built F-104S Starfighters was ordered by the AMI and have been delivered, and these equip three interceptor wings or Stormi (No 4 Wing at Grosseto, No 9 at Grazzanise, No 53 at Cameri), each with a squadron or Gruppo of 12 aircraft, Nos 9, 10 and 21 Squadrons respectively. Three further wings have a strike/interceptor role with the F-104S: No 5 Wing at Rimini-Miramare with Nos 23 and 102 Squadrons, No 36 Wing at Gioia del Colle with Nos 12 and 156 Squadrons and No 51 Wing at Istrana (Treviso) with No 22 and 155 Squadrons. No 6 Wing at Ghedi,

Below: *Aeritalia G91Y of No 32 Wing, Brindisi, Italian air arm.* / State Maggiore Aeronautica

Above: *Italian Navy AS-61ASW fully equipped with torpedoes and missiles.*

comprising No 154 Squadron, operates F-104Gs in the bomber role, while No 3 Wing at Verona-Villafranca, with Nos 28 and 132 Squadron flies F-104Ss and some 30 RF-104Gs in the tactical reconnaissance role. Italy purchased 125 F-104Gs from the large European production programme in the 1960s and of those withdrawn from use 20 have been transferred to the Turkish Air Force. To supplement the F-104S units, 100 Panavia Tornado variable-geometry strike aircraft are to join the AMI made up of 54 for front-line use with Nos 20, 102, 154 and 186 Squadrons, 12 dual control trainers and 34 in reserve.

A fighter-bomber wing, No 8 at Cervia-San Giorgio with No 101 Squadron, operates the twin-engined Aeritalia G91Y, together with No 13 Squadron attached to No 32 Wing at Brindisi. A total of 65 Ys were purchased before the production line closed. The AMI's remaining G91R strike/reconnaissance element comprises No 2 Wing at Treviso-San Angelo with Nos 14 and 103 Squadron. The official Italian aerobatic team, Frecce Tricolori, has G91PANs in a dual combat/demonstration role as No 313 Squadron at Rivolto. The team is to replace it aircraft with the new MB339 during 1979.

Italy's fixed-wing anti-submarine units come under the command of the Navy or Marinavia, but are retained on the Air Force establishment. At Cagliari Elmas, No 30 Wing with No 86 Squadron has Breguet Atlantics, while at Catania, No 41 Wing operates Nos 87 and 88 Squadrons with 18 Grumman S-2F Trackers and the remainder of the 18 Atlantics procured in 1972.

Transport and Support

Transport duties are performed by the 46th Brigade at Pisa, made up of No 2 Squadron with 20 Fairchild C-119Gs, No 50 Squadron with 13 Lockheed C-130H Hercules and No 98 Squadron with Aeritalia G222s (15 aricraft). A total of 44 G222 STOL transports are on order for the AMI made up of 38 freighters, four calibration aircraft and two ECM variants. An SAR wing, No 15 at Ciampino with Nos 84 and 85 Squadrons, is equipped with 12 Grumman HU-16A Albatrosses in the former unit and seven Agusta-Bell AB47Js and 14 AB204s in the latter. Replacing the helicopters are 20 Agusta-built Sikorsky HH-3Fs, delivery of which began early in 1978.

No 14 Wing at Pratica di Mare near Rome is an ECM and radio calibration unit formed from No 8 and 71 Squadrons, and equipped with four Piaggio PD808s, two EC-47s, three EC-119s and some T-33As. The G222 specialised versions are due to replace the EC-119s by 1979. Based at Ciampino are Nos 92 and 306 Squadrons forming No 31 Wing for VIP and Government use, the former squadron havng only two Agusta-built Sikorsky AS-61TS helicopters and the latter unit flying two McDonnell Douglas DC-9-30s, two DC-6s and some PD808s. A total of 25 PD808s are in AMI service plus 90 AB47Js, 60 AB204Bs and 60 AB206s. Three regional communications units are divided as follows: 1 Region at Milan flying P166Ms, T-6s, S208Ms, AB47Js and AB204Bs; 2 Region at Guidonia flying T-6s, P166Ms, S208Ms and C-47s;

3 Region at Bari-Palese with AB204Bs and P166Ms. There are 51 Piaggio P166Ms and 44 Siai-Marchetti S208Ms in use.

Training begins on the Siai-Marchetti SF260AM, 20 of which equip No 207 Squadron at Latina, students graduating to the Aermacchi MB326 at the basic training school at Lecce-Galatina. A total of 130 MB326s fly with Nos 212, 213 and 214 Squadrons at the school and are scheduled for replacement by the new MB339, of which 100 are on order. Advanced training is conducted at Foggia-Amendola on approximately 100 two-seat G91Ts equipping Nos 201, 204 and 205 Squadron while 28 two-seat TF-104Gs of No 20 Squadron at Grosseto have an operational conversion role.

The Italian Navy air component, known as the Marinavia, is primarily a helicopter force, shore-based at Catania and Luni (La Spezia) with detachments at other ports with a primary function of supplying Italian naval vessels with ASW air elements. A total of 24 Agusta-built Sikorsky SH-3D Sea Kings equip No 1 Squadron at Luni and No 3 Squadron at Catania and assigned to cruisers and the projected 10,000 ton helicopter carrier *Garibaldi* which will have accommodation for up to 16 Sea

Above: *Sixty Agusta A109s have been ordered for the Italian Army.*

Kings; in-service date for the vessel is 1982. Two further squadrons, Nos 2 Squadron at Catania and No 5 Squadron at Luni, have 30 AB204AS for use aboard smaller ships but being steadily replaced by 42 AB212ASW helicopters ordered between 1975 and 1977; at least 23 had been delivered by mid-1978. Five A106 light torpedo-carrying helicopters operate from Impavido-class warships. For training duties and liaison work there are 12 AB47J.

Known as the Aviazione Leggera dell'Escercito (ALE), the Italian army air component has more than 400 machines comprising 100 AB47G/Js, 50 AB204Bs, 138 AB205s, and 142 AB206s ordered for light observation duties. For heavy-lift work there are 25 of 26 Meridionali-built Boeing CH-47C Chinooks delivered, together with five A109 Hirundos for TOW missile trials; an order for 60 has been placed with Agusta for anti-tank use. Fixed-wing aircraft include about 110 Cessna 0-1Es and Piper Super Cubs, and more than half of the 80 SM1019s on order for air observation post duties to replace the older types.

Ivory Coast

Title: Force Aérienne de Côte d'Ivoire
Headquarters: Abidjan
Strength: 22 aircraft

Below: *One of the three Douglas C-47s of the Ivory Coast Air Force.* / H. Holmes

The air component of this ex-French colony operates in support of the Ivory Coast Army, flying liaison and transport duties. However, the acquisition of 12 Alpha Jet trainers announced in 1977, heralds the formation of a small dual training/counter-insurgency unit when the aircraft are delivered in 1979-80. For transport duties there are two F28 Mk 1000s, three

Douglas C-47s, a Fokker F27 Mk 400M, three Reims F337s, three SA330 Puma helicopters, three Alouette IIIs and two Alouette IIs. At Abidjan, the main air force base, is stationed the VIP/Government flight equipped with a Dassault Falcon 20, one Fokker F27 Mk 600, one F28 Mk 4000 and one Aero Commander 500B. For basic training there are two Reims-Cessna 150s.

Jamaica

Title: Jamaica Defence Force Air Wing
Headquarters: Kingston
Strength: 14 aircraft

A transport and liaison force with additional SAR and police co-operation roles, the Air Wing is based at Up Park Camp, Kingston, and operates a variety of types purchased from western sources. The fixed-wing element has two Britten-Norman Islanders for utility work, a single DHC Twin Otter 300 delivered in 1967 for VIP flights, a Beech Duke received in 1975, two Cessna 185B Skywagons and a Beech King Air 100 for transport duties. The helicopter flight operates three Bell 212 heavy-lift machines and four Bell JetRangers for liaison.

Below: *DHC Twin Otter 300 of the Jamaican Defence Force.* / DHC

Japan

Title: Japan Air Self-Defence Force
Headquarters: Tokyo
Strength: 900 aircraft

Combat
Japan is divided into three regional air commands for air defence purposes, Northern, Central and Western, plus the Southwest Air Wing in Okinawa, all linked to the Base Air Defense Ground Environment (Badge) system. There are seven wings — Nos 2, 3, 5, 6, 7, 8, Southwest and General HQ — operating a total of 14 squadrons. The 169 Mitsubishi-built Lockheed F-104J Starfighters equip six units, Nos 202-207 Squadrons flying in the interceptor role, supported by a steadily diminishing number of North American F-86F Sabres, flying as advanced trainers in three squadrons (Nos 6, 8 and General HQ) and the 1st Air Wing of the Air Training Command; most of the surviving 180-odd Sabres will be retired by 1980.

Five McDonnell Douglas F-4EJ Phantom squadrons totalling 140 aircraft are planned within five air wings, Nos 2, 6, 7, 8 and the Southwest, of which four had formed by mid-1978 (Nos 301-304 Squadrons); the fifth, No 305, due to be operational by the beginning of 1979, replacing No 206 (F-104J) Squadron. A reconnaissance unit, No 501 Squadron, is equipped with 14 RF-4EJ Phantoms, a further 14 being planned for procurement.

In 1977, JASDF selected the McDonnell Douglas F-15C/D Eagle as a replacement for the Starfighters and after cost problems had been resolved ordered 100 aircraft to be licence-built by Mitsubishi between 1980 and 1988 for four squadrons; a further 23 F-15s may be purchased for a fifth squadron when production is under way. To replace the Sabres, the Mitsubishi-designed F-1 close-support fighter is being produced to meet orders for 59 aircraft. Several more will be needed to equip a total of three squadrons, the first unit having already formed, No 3 Squadron of the 3rd Air Wing, in March

Top: *The Japanese Air Self-Defence Force's Starfighters are being replaced by F-15s in the 1980s.* / H. Holmes

Above: *Japanese Ground Self-Defence Force's Bell UH-1B.*

1978. The JASDF is to receive four Grumman E-2C Hawkeye early warning aircraft in the Fiscal 1979 budget.

Transport and Support
The Transport Wing of the JASDF comprises three squadrons, Nos 1, 2 and 3, equipped with 26 Kawasaki C-1 twin-jet transports replacing Curtiss C-46Ds, and a dozen NAMC YS-11As. There is an ECM training unit with a modified YS-11E and two T-33As, and a flight check unit with two YS-11s, three MU-2Js and four T-33As. The Air Rescue Wing operates 21 Mitsubishi MU-2Ss, 29 Kawasaki-built Boeing KV-107-IIA helicopters and seven S-62As, with two more KV-107s due for delivery in 1979.

Five training wings are in operation: No 1 with some 50 F-86Fs and about 50 T-33As for advanced training, No 4 with 46 Mitsubishi T-2As, No 11 and 12 with some 80 Beech T-34As for primary training; and No 13 Wing, with 50 T-1A/Bs for intermediate duties. The Sabres in the 1st Wing are due for retirement in 1978 and 66 T-2As are on order with deliveries due for completion by March 1981. The first 44 of 60 Fuji T-3s to replace the T-34As in Nos 11 and 12 Wings have been ordered.

Title: Japan Maritime Self-Defence Force
Headquarters: Shimofusa
Strength: 311 aircraft

The JMSDF anti-submarine force comprises five shore-based air groups plus an independent unit assigned to the nation's Defence Fleet, and three independent smaller units controlled by each district command. Three (Nos 1, 2 and 4) of the five groups and the independent unit in Okinawa are equipped with 77 Kawasaki-built Lockheed P-2J Neptunes, 13 Lockheed P-2Hs and 28 Grumman S-2A Trackers, while the fourth group (21st) and the other two independent units are operating 69 Sikorsky SH-3A Sea Kings and the fifth group (31st) is flying 17 Shin Meiwa PS-1 amphibians. Of the 20 PS-1s delivered by mid-1978, four have been lost in accidents. A further six P-2Js, three PS-1s and 14 SH-3As are planned for delivery to the JMSDF before March 1980. To replace the P-2J/H force, 45 Lockheed P-3C Orions are being purchased by the service between 1981 and 1988. Kawasaki will build them and four squadrons will be formed beginning in 1982.

A helicopter-equipped mine-sweeping unit, No 111 Squadron, has 11 KV-107-IIAs and is expected to receive two Sikorsky RH-53Ds if the Fiscal 1979 request is approved. An air rescue unit, No 71 Squadron, has three Shin Meiwa US-1 amphibians, three S-61As (used by the South Pole Observatory Group) and eight S-62As. Three more US-1s on order for the squadron are due to be delivered before 1980. No 61 Squadron provides transport facilities for the JMSDF flying four YS-11s and an S-2A Tracker. Four training groups have six YS-11Ts, five King Air 90s and 28 Beech Queen Airs for instrument work, 26 KM-2s for intermediate training; a helicopter training element has three Hughes OH-6Js and eight Bell 47s.

Title: Japan Ground Self-Defence Force
Headquarters: Tokyo
Strength: 413 aircraft

Nineteen aircraft are being purchased by the Army's air component in the Fiscal 1979 budget. These comprise three Mitsubishi LR-1 liaison and reconnaissance aircraft, a Kawasaki KV-107-IIA, three Bell UH-1Hs, 12 OH-6Ds and one Bell AH-1S anti-tank helicopter. JGSDF is organised into 24 squadrons operating at divisional level through five army commands covering the whole of Japan and Okinawa. The current fixed-wing force comprises 13 LM-1/2s, 27 Cessna O-1A/Es, and eight LR-1s plus two T-34A trainers. The LM-1/2s and O-1s are being steadily withdrawn in favour of helicopters. This large rotary-wing force is planned to total 56 KV-107-IIAs, 54 UH-1Hs, 117 OH-6Js and 38 TH-55Js. In addition there are 82 Bell UH-1Bs in service for transport duties. Two Bell AH-15 anti-tank helicopters have been bought for evaluation prior to a possible order for 32, and the JGSDF has selected the Boeing Chinook as a replacement for the KV-107s, with up to 40 being required.

Jordan

Title: Royal Jordanian Air Force
Headquarters: Amman
Strength: 138 aircraft

Combat

In the future RJAF expansion programme, the front-line force will be increased to 176 fixed-wing aircraft, the Northrop F-5 element will be enlarged and the F-104 Starfighters will be withdrawn and replaced by a new, as yet unspecified, interceptor. After protracted negotiations, a US air-defence system was ordered in 1976, involving Improved Hawk anti-aircraft missiles, Vulcan guns and Redeye AA infantry missiles. The original Northrop F-5E order for Jordan covered 30 aircraft, but this was subsequently increased to 57 F-5E interceptors and six F-5F two-seat trainers. No 17 Squadron at Prince Hassan air base received the initial batch of 24 aircraft, and the others are joining Nos 1 and 2 Squadrons at King Hussein air base, Mafraq, which until recently has operated 30 ex-Iranian F-5As on fighter-bomber duties. Four F-5B trainers have provided two-seat conversion for all three units. Also at Prince Hassan AB is No 9 Squadron which is equipped with 18 Lockheed F-104A interceptors and four two-seat F-104B conversion trainers.

Transport and Support

Based at King Abdullah air base, Amman, is the modest transport force of four Lockheed C-130H Hercules (replacing four C-130Bs, two of which have been transferred to the Singapore AF) delivered in 1978, and four Spanish Casa C212 Aviocar utility freighters; one of the Aviocars has an executive interior for government duties. A total of 15 Aérospatiale Alouette IIIs are based at Amman with detachments assigned to other bases for SAR duties. Four Sikorsky S-76s are tentatively on order but confirmation is still awaited, as is the report that 12 SA342K Gazelles are due for delivery in 1978-79.

At the Royal Jordanian Air Academy in Amman, 12 Scottish Aviation Bulldog elementary trainers are used for intitial flying training, students then transferring to No 6 Squadron at Mafraq which operates 12 ex-USAF Cessna T-37Cs in the advanced role. At the personal disposal of King Hussein are a Riley Dove and a Boeing 727 airliner.

Below: *Jordanian F-104A interceptor of No 9 Squadron.*

Kampuchea

Title: Khmer Liberation Army
Headquarters: Phnom-Penh
Strength: Unknown

Fierce border clashes with neighbouring Vietnam culminated in the invasion of Kampuchea in January 1979 and the swift destruction of the country's armed forces. China had supplied Kampuchea with some 16 Shenyang F-6s (MiG-19s) but only six of these were uncrated at the airfield of Kanpong Chnang, 50 miles north-east of the capital. A further 10 were captured by Vietnamese forces, still crated. Types known to have been in the former Cambodian Air Force at the time of the Communist take-over in spring 1975 include Douglas AC-47 Gunships, Fairchild C-123 Providers and C-47s, Cessna O-1 observation aircraft, Bell UH-1H helicopters and North American T-28D attack-trainers, totalling some 100 aircraft.

Kenya

Title: Kenya Air Force
Headquarters: Nairobi
Strength: 86 aircraft

Combat

Formed with British assistance following Kenya's attainment of independence in 1963, the Kenya Air Force operates a front-line fighter-bomber force of 10 Northrop F-5E Tiger IIs and two F-5F two-seat combat trainers delivered in 1978. Based at Nanyuki alongside the F-5 squadron is a unit with five Hawker Siddeley Hunter FGA9s which were received in 1974; a sixth Hunter was lost subsequently. A further ground-attack squadron is likely to be formed with 12 British Aerospace Hawk MK 52 strike/trainers which were unofficially ordered in 1978. Also in service is an attack/trainer unit with five of six BAC Mk 87 Strikemasters originally received in 1971 and the country's first jet equipment.

Transport and Support

Canadian aircraft predominate in the two transport squadrons of the KAF based at Nanyuki, six de Havilland Canada DHC-5D Buffaloes and eight DHC Caribous providing medium-lift duties and flying supply missions to the many small airstrips in remote areas of the country. Light transport work and locust control is the task of 15 DHC Beavers while a more recent acquisition is a batch of six Dornier Do28 Skyservant light STOL freighters. For VIP and liaison flights there is an Aero Commander 680F and a Piper Navajo Chieftain.

Scottish Aviation Bulldog 103 trainers are in use in the primary role, a total of 14 having been purchased in two batches of five and nine aircraft, with students then proceeding to the Strikemasters for advanced flying. Helicopters include two Bell 47Gs, two SA342K Gazelles, at least three Aérospatiale Alouette IIs and reportedly, some six French-supplied SA330 Puma tactical assault machines.

Below: Kenyan Air Force Dornier Do28 Skyservant.

Korea (North)

Title: Korean People's Army Air Force
Headquarters: Pyongyang
Strength: 850 aircraft

Combat

North Korea relies heavily on military and economic aid from Russia and China and consequently operates a wholly Eastern Bloc array of aircraft. There are currently nine squadrons flying approximately 170 Soviet-supplied MiG-21F/PF interceptors supplemented by an additional force of some 100 MiG-19s. The establishment of a MiG-21 assembly line in the country was reported in the mid-1970s but no further confirmation has been obtained. A large ground-attack element of more than a dozen squadrons has about 350 MiG-17s and more than 30 Su-7Bs; a bomber force has 70 Il-28s.

Transport and Support

There are only about 40 transport aircraft in use with KPAAF made up of An-2s and Il-14s, while a VIP unit has two Il-18s and a Tu-154B. A helicopter force totals about 20 Mi-4s and a similar number of Mi-8s. Seventy Yak-18s and MiG-15UTIs, and a few two-seat MiG-21Us, are used for training.

Korea (South)

Title: Republic of Korea Air Force
Headquarters: Yong Dong Po City
Strength: 500+ aircraft

Combat

This major Far Eastern air force is firmly based on American military support and organisation, and a steady modernisation programme is being conducted to help strengthen it. As well as having had major military overhaul facilities in the country for many years, an assembly line has been established for the Hughes 500M-D Defender multi-role helicopter. Equipped with TOW missiles for the anti-tank role, 100 Defenders are being delivered to the South Korean armed forces, particularly the Army. The current RoKAF combat force comprises two all-weather fighter squadrons equipped with 18 McDonnell Douglas F-4D Phantoms originally supplied by the USA in 1969, and 19 F-4Es supplied in the early 1970s; a further 18 Es are scheduled to be delivered for a third squadron during 1979. Nearing completion is an order for 126 Northrop F-5E fighter-bombers and 20 two-seat F-5Fs for four squadrons, replacing on a one-for-one basis the older force of 87 F-5As and 35 F-5Bs. A reconnaissance unit has 12 RF-5As and there are two remaining squadrons of North American F-86F Sabres totalling some 50 aircraft, of 112 originally received.

Other new types on order or under active consideration, include the Rockwell OV-10G Bronco counter-insurgency aircraft of which 24 are being delivered following a 1976 US Military Sales authorisation; the General Dynamics F-16 lightweight fighter for which a request for 72 has been made; and the Fairchild A-10A Thunderbolt II for which the figure of 50 aircraft has been mentioned.

Below: *Bell 212 for the South Korean air arm.*

Transport and Support

To increase the effectiveness of the transport force, six Lockheed C-130H Hercules are being delivered to replace older machines which have seen many years service. Forty Curtiss C-46s, Douglas C-54s, Fairchild C-123Ks and Aero Commanders constitute the present Air Transport Group which also operates two HS748s for Presidential use. The helicopter force has six Sikorsky H-19s, five Bell UH-1Ds, one UH-1N (for VIP use) and two Bell 212s, these being used chiefly for SAR duties. Cessna O-1s and 12 Cessna O-2As fly forward-air-control duties, while liaison flying is conducted by Cessna U-17s and some DHC Beavers. The training elements are equipped with 20 Cessna T-41Ds for primary tuition, 24 North American T-28Ds for basic instruction, and 30 Lockheed T-33As for advanced work; the two-seat Northrop F-5B/Fs are tasked with operational conversion flying.

The South Korean Navy has a shore-based ASW unit equipped with 20 Grumman S-2A/F Trackers delivered from 1970 onwards, and is to receive some of the Hughes 500M-D Defenders which are being assembled in the country. The RoK Army is also getting Defenders to supplement the Bell UH-1Ns it currently operates; other types in use include Cessna O-1s, DHC Beavers and Hiller OH-23s.

Kuwait

Title: Kuwait Air Force
Headquarters: Kuwait
Strength: 110 aircraft

Combat

To protect the oilfields of this rich Arab state, all three Kuwaiti armed forces are being strengthened and new equipment phased into service. The Kuwait Air Force has received a large share of the military budget over the past few years and now has three front-line combat squadrons in service with modern equipment. For strike duties there are two squadrons of McDonnell Douglas Skyhawks — 30 A-4KU single-seaters and six TA-4KU two-seat trainers — based at one of two new airfields constructed by Yugoslav contractors to the south and west of the KAF's main fighter base at Kuwait. Also in service are 18 Dassault Mirage F1CK strike fighters and two F1BK two-seat combat trainers in a single interceptor squadron which had previously been equipped with 10 BAC F53 Lightnings and two T67 Lightning trainers.

A ground attack/trainer squadron of nine BAC Mk 83 Strikemasters is in service of 12 originally delivered in 1969 and 1971. Five HS Hunter T67s complement these, but the four single-seat FGA57s have been retired from use.

Transport and Support

To meet a pressing long-range transport requirement, two McDonnell Douglas DC-9-30 convertible passenger/cargo aircraft were bought in 1976 and operate alongside two Lockheed L-100-20 freighters delivered in 1971. The helicopter force has been enlarged from a handful of Italian types to two squadrons of Aérospatiale SA342 Gazelles totalling 24 machines. One unit has an LOH role and the other an anti-tank mission for which its Gazelles are fitted with HOT missiles. A third squadron has a troop carrier role flying 12 SA330 Puma assault helicopters.

Below: *One of the two Kuwait Air Force Dassault Mirage F1BK trainers.* / AMD-BA

Laos

Title: Air Force of the Liberation Army
Headquarters: Vientiane
Strength: Unknown

Known before the communist takeover in April 1975 as the Royal Lao Air Force, the AFLA's equipment status remains obscure with spares for the surviving American aircraft becoming difficult to obtain, thus reducing the effectiveness of the arm as a whole. However, help has been received from the Soviet Union when, in 1977, 10 MiG-21s were delivered to Vientiane's Wattay airport accompanied by Soviet technicians. From the same source came small batches of An-24 transports, An-2 utility biplanes and Mi-8 helicopters.

American aircraft remaining in the country after the takeover included 63 North American T-28Ds and 10 Douglas AC-47 Gunships for attack duties, four Cessna U-17As, one Aero Commander 500 and a single DHC Beaver. The transport force had 18 Douglas C-47s, 42 Sikorsky UH-34s and six Alouette II/IIIs, the helicopters being used for rescue and Army support work. Six Cessna T-41Ds formed a primary training unit.

Lebanon

Title: Force Aérienne Libanaise
Headquarters: Beirut
Strength: 68 aircraft

During the tragic Lebanese civil war in the late 1970s, the Air Force conducted low-key operations, consisting mainly of liaison and communications duties with helicopters. The combat elements rely on 17 HS Hunter F70 fighter-bombers and two T66 two-seat trainers for operational duties, with the single squadron of 10 Dassault Mirage IIIEL interceptors and one IIIBL trainer apparently in storage following their delivery in 1965. A DH Dove 6 undertakes transport flights alongside four Alouette IIs, 13 Alouette IIIs and six Agusta-Bell AB212s. For basic training six Scottish Aviation Bulldog 126s were received in 1975, while advanced tuition is performed on eight Fouga Super Magisters.

Below: *Lebanese Air Force HS Hunter F70.*

Liberia

Title: Liberian Army/Air Reconnaissance Unit
Headquarters: Monrovia
Strength: 8+ aircraft

This airborne element of the 5,000-man Liberian Army was established in the early 1970s for liaison and communications duties. The largest aircraft in service is the Douglas C-47, of which there are two for transport flights, supplemented by a few Cessna 337s bought in 1978, one Cessna 185 and one Cessna 207. Two Cessna 172s are used for training.

Libya

Title: Libyan Arab Republic Air Force
Headquarters: Tripoli
Strength: Unknown

Combat

One of the few remaining Arab countries still receiving large-scale military and economic aid from the Soviet Union. Libya continues to expand her armed forces with modern equipment purchased from France, Italy and Russia. Following a substantial arms deal signed in the mid-seventies between Libya and Russia, the Libyan Arab Republic Air Force has received two squadrons of MiG-23 interceptors, MiG-27 strike aircraft and a few two-seat MiG-23Us totalling some 50 aircraft, a few Tu-16 medium bombers and a squadron of 12 Tu-22 strategic reconnaissance aircraft. A large number of Soviet Air Force personnel man and maintain many of these aircraft alongside the few trained Libyans, and as part of the arms deal Russia maintains a MiG-25 reconnaissance unit in the country for surveillance flights over the Mediterranean.

From France Libya continues to receive Dassault Mirages following orders placed through the 1970s for an initial order for 110 followed by a further 38. The former purchase covered 60 Mirage 5D fighter-bombers for two squadrons and an OTU, 30 5DE interceptors in two further squadrons, 10 5DD two-seat trainers and 10 5DR reconnaissance aircraft. Joining this force in 1978 are 38 Mirage F1s (plus 50 on option) comprising 16 F1AD strike aircraft, 16 F1ED interceptors and six F1BD trainers.

Transport and Support

The bulk of the transport arm operates within two squadrons equipped with eight Lockheed C-130H Hercules delivered in 1971 (with a further eight embargoed in the United States), nine Douglas C-47s, two Dassault Falcon 20s and a Lockheed JetStar for VIP use. The American embargo on spares for the Hercules in causing the Libyans major problems and to partially solve the difficulty, an order for 20 Rolls-Royce Tyne-powered Aeritalia G222s was placed in 1978 for delivery in 1980. Helicopters have been bought from France and Italy, while Russia has supplied about 12 Mi-8s. Boeing-Vertol CH-47C Chinooks, built in Italy by Meridionali total eight in LARAF service with another eight on order in 1978. Also in use are nine Aérospatiale SA321M Super Frelons for heavy-lift, SAR and ASW, 10 Alouette IIIs, three Alouette IIs, three Bell 47Gs, two AB212s and a single Agusta-built AS-61A-4 for VIP duties. A single Super Frelon and three of the Alouette IIIs have been donated to Malta for SAR duties.

Twelve ex-French AF Fouga Magisters equip a training unit together with three Lockheed T-33As and two Dassault Mystere 20 Mirage radar trainers. To update the training elements about 38 Yugoslav Galebs have been delivered recently and Italy is benefiting from a substantial order for 250 Siai-Marchetti SF260WL primary trainers; 20 have already been delivered and the Italian company is setting up an assembly plant in Libya for production of the aircraft.

The Libyan Army operates a small air section equipped with five AB206 JetRangers, four Alouette IIIs and six AB47Gs, together with a few ex-Italian AF Cessna 0-1s for liaison duties.

Madagascar

Title: Armée de l'Air Madagascar
Headquarters: Ivato
Strength: 19 aircraft

Previously known as Malagasy, this island republic has reverted to the name used when it was a French colony. A small transport air arm was established by the French in the early 1960s and current equipment still reflects this aid. One Douglas C-53D and five C-47s are the largest aircraft in service being used for transport work, supplemented by one Piper Aztec D, three Reims Super Skymasters and a Britten-Norman Defender, the latter also being flown on policing and counter-insurgency duties. For training there are four Cessna 172Ms while the helicopter element consists of one Bell 47G, one Alouette II and two Alouette IIIs.

Malawi

Title: Malawi Air Wing
Headquarters: Blantyre
Strength: 10 aircraft

This small air component is part of the Malawi Army and operates from Lilongwe, the main operational base. Transport, supply and liaison flights are undertaken with four Douglas C-47s and six Dornier Do28D-2 Skyservants, the first two of which were delivered early in 1976. An SA330L Puma and an Alouette III have also been purchased.

Malaysia

Title: Royal Malaysian Air Force
Headquarters: Kuala Lumpur
Strength: 200 aircraft

Combat

Units of the RMAF have been operating against Communist guerrillas along the Malaysia-Thailand border for a number of years in efforts to reduce infiltration and activity by these dissidents. It is mainly a helicopter operation but a ground-attack squadron, No 6, has flown strike sorties using a dozen Canadair CL-41G Tebuan armed trainers. A total of 16 of these aircraft are in service with the RMAF (of 20 delivered), the remainder based at Kuantan with No 9 Squadron in the advanced trainer role. For air defence, the Service has two squadrons of Northrop F-5E/B interceptors based at Butterworth — No 11 'Cobra' Squadron, previously flying Commonwealth Sabres, and No 12 'Tiger' Squadron — the 14 F-5Es and two F-5B trainers being delivered in 1975-76.

Also based at Butterworth are two Royal Australian Air Force Mirage interceptor squadrons, No 3 and 75, which form part of the five-power agreement involving Malaysia, Australia, New Zealand, Singapore and the UK.

Transport and Support

Long-range supply flights are flown by No 14 Squadron's six Lockheed C-130H Hercules received in 1976 and these are supported by Nos 1 and 8 Squadrons at Labuan equipped with 17 DHC Caribou STOL transports delivered in batches between 1965 and 1973. With the Hercules at Kuala Lumpur is No 2 Squadron, the VIP/Government unit operating two Fokker F28 Mk 1000s, two HS125s, three DH Herons and two Doves. Liaison and forward air control duties are performed by two helicopter squadrons with 25 Aérospatiale Alouette IIIs; No 3 Squadron at Kuala Lumpur and No 5 Squadron at Labuan. Heavy-lift and troop transport capacity is provided by 36 Sikorsky S-61As flying with No 7 Squadron at Kuching and No 10 Squadron at Kuantan. Known as Nuris in RMAF service, the S-61s have been in use for some years.

The RMAF Flying Training School at Alor Star in West Malaysia has 15 Scottish Aviation Bulldog 102s for primary training and 12 Cessna 402Bs delivered in 1975 for multi-engined training and survey work. A helicopter FTS is situated at Labuan equipped with six Bell 47Gs and three ex-British Army Sioux. Five Bell 206Bs and three Bell 212s have been received in more recent years, while from France, 20 Aérospatiale SA341K Gazelles are being delivered for a variety of duties.

Below: *Northrop F-5E of the Royal Malaysian Air Force.*

Mali Republic

Title: Force Aérienne du Mali
Headquarters: Bamako
Strength: 12+ aircraft

A former French colony, Mali has been courted by the Soviet Union and operates a token combat force of five MiG-17 fighter-bombers and one two-seat MiG-15UTI trainer supplied some years ago. For transport duties there are two Mi-4 helicopters, two An-2 biplane light utility aircraft, two Douglas C-47s and some Yak-18 trainers.

Malta

Title: Armed Forces of Malta
Headquarters: Sliemma
Strength: 9 aircraft

Headquartered at St Patrick's Barracks, Malta, this small air arm comprises a helicopter flight equipped with four Bell 47Gs originally donated by West Germany in 1971 and an AB206 presented by Libya in 1973. Duties include general surveillance, transport and anti-smuggling patrols. Two or three ex-West German Do27 liaison aircraft were expected to join the Force but delivery had not been effected by mid-1978. Following the withdrawal of British forces in 1978, Libya pledged further aid to Malta and tangible evidence included the supply of three Alouette IIIs and a Super Frelon from Libyan Arab Air Force stocks.

Mauretania

Title: Mauretania Islamic Air Force
Headquarters: Nouakchott
Strength: 19 aircraft

A limited modernisation programme for this small air arm began in 1975 in an effort to combat the increasing activities of the Polisario rebels in the former Spanish Sahara. Initially a transport arm, the Air Force subsequently acquired a small attack and counter-insurgency element in the form of six Britten-Norman Defenders delivered in 1976-77. Two were later lost on operations and a further three were ordered late in 1977 to make up for attrition.

To expand the transport and logistical capabilities of the arm, two DHC Buffaloes were ordered in 1977, but prior to the delivery of these machines, two Canadian Armed Forces Buffaloes were loaned to MIAF in 1978 being based at Nouakchott alongside the two Short Skyvan 3Ms received in 1975 and flown jointly by the army and air force. A single Caravelle is used for VIP flights and two Douglas DC-4s conduct long-range transport flights. Also in service are two MH1521M Broussards for liaison and four Reims F337 Super Skymasters flown on FAC, training and Coin duties. An order for Argentinian Pucara turboprop twins foundered in 1978 due to financial problems.

Below: *Four Rheims F337 Super Skymasters fly with the Mauretanian Air Force.* / S. R. P. Thomson

Mexico

Title: Fuerza Aerea Mexicana
Headquarters: Lomas de Sotelo
Strength: 258 aircraft

Combat

The lack of finance and Mexico's comparatively harmonious accord with her neighbours are factors which have governed the FAM's apparent reluctance to procure modern combat aircraft. At one stage, Israeli Kfir production was mooted as was the possibility of Mexico's acquisition of Northrop F-5Es, but both moves were subsequently abandoned. Instead the air force soldiers on with a strike/trainer squadron, No 202 at Santa Lucia AB, equipped with 15 Lockheed T-33As, supported by 50 North American T-28As in Nos 201, 205, 206 and 207 Squadrons, and some 20 North American T-6s in Nos 203 and 204 Squadrons at El Cipres AB, although not all the aircraft are operational. A light support unit at Cozumel AB, No 209 Squadron, uses 18 LASA-60s mainly for SAR duties, plus one Hiller 12E and nine Alouette IIIs.

Above: *Mexican Air Force Douglas C-47.* / G. Pennick

Transport and Support

A Heavy Transport Squadron is based at Santa Lucia equipped with one Douglas DC-7, five C-54s and two C-118s, while a Light Transport Group operates seven Douglas C-47s, one Short Skyvan, 20 Aero Commander 500s, 12 Britten-Norman Islanders (Government-operated) and 10 IAI Aravas. Based at Mexico City is a VIP and government squadron flying two Boeing 727-100QCs, an HS125, a BAC One-Eleven and a Lockheed JetStar. The helicopter force stands at 14 Bell 47Gs, five Bell JetRangers, a Bell 212 and about 10 Bell 205s. Twenty aerobatic Bonanza F33Cs are in service for training, joining 20 Beech Musketeer Sports bought in 1970, and some Beech B55 Barons used for instrument training. The FAM is to receive 12 Pilatus PC-7 Turbo-Trainers to update the training syllabus at the basic stage, and these aircraft are expected to have provision for underwing armament, thus doubling in the Coin role; delivery is due in 1979.

The Mexican Navy operates four Grumman HU-16A Albatrosses for ASW and SAR duties, and another four have recently been delivered from ex-US Navy stocks. At least two C-47s, four Alouette IIs and five Bell 47Gs are on strength. Two Bonanzas have been delivered and a VIP Learjet 24D is in service alongside a Fairchild FH-227.

Mongolia

Title: Air Force of the Mongolian People's Republic
Headquarters: Ulan Bator
Strength: 60 aircraft

This Soviet satellite country has an Army-controlled air force with an inventory almost unchanged for more than a decade. There is one ground-attack unit with 10 MiG-15/17s, a transport element with about 30 aircraft made up of An-2s, An-24s and Il-14s, and a helicopter unit with 10 Mi-1 and Mi-4 machines. Trainers include a few Yak-11s and Yak-18s. It is conceivable that a few more modern aircraft of Soviet origin are in service but confirmation of this is still awaited. However, more recently three PZL-104s have reportedly been received from Poland.

Morocco

Title: Forces Armées Royales
Headquarters: Rabat
Strength: 250+ aircraft

Combat

A major re-equipment programme is currently underway, which will significantly increase the offensive potential of the Royal Moroccan Air Force through the 1980s. Most important new type to be ordered is the Dassault Mirage F1CH of which there are 50 due for delivery beginning in 1979. The purchase was agreed in 1975 and, according to the French manufacturers, Dassault, only the single-seater version is involved, although an option on a further 25 aircraft remains open and could see some two-seat trainers included if converted to a firm order. While the Mirages are expected to be assigned the interceptor role, 24 French Alpha Jet Es ordered in 1978 will have a dual task, that of strike/trainer. Delivery will be in 1980. Morocco's request in 1978 for 24 Rockwell OV-10 Bronco counter-insurgency aircraft and some Bell AH-1 helicopter gunships from the USA was refused on political grounds.

Since 1966, Morocco has operated Northrop F-5As and at present there are two squadrons with 14 aircraft, of 18 delivered plus two RF-5As, with three two-seat F-5Bs acquired at the same time for conversion training. Two further ground-attack squadrons have 24 Fouga Magisters and all four units were heavily involved in the recent guerrilla war in the Western Sahara. In storage are 12 MiG-17s and two MiG-15UTI trainers supplied by the Soviet Union in 1961.

Above: *First of 12 Beech T-34Cs for Morocco delivered in 1978.* / Beechcraft

Transport and Support

In 1973, six Lockheed C-130H Hercules were ordered for transport duties, replacing a fleet of 18 Fairchild C-119G Packets originally supplied by the US, Canada and Italy. A second batch of six Hercules arrived in the country during 1977 and re//placed 10 Douglas C-47s. For staff and liaison flights, six Beech King Air 100s are in use together with a dozen MH1521M Broussards supplied by France in the 1950s. Helicopters include 24 Agusta-Bell AB205s delivered in 1970, at least eight AB206s, five AB212s, four Kaman HH-43B Huskies, four Alouette IIs, four Bell 47Gs, six SA342 Gazelles and 40 SA330 Pumas. For heavy lift duties six Meridionali CH-47C Chinook helicopters were ordered in 1978 for delivery in 1980; a repeat order is likely.

The training elements are phasing in several new types after many years equipped with North American T-6s (about 25 remain in use) and T-28s (15 disposed of in 1978) supplied by France in the 1960s. For primary tuition, 10 Swiss AS202 Bravos were purchased in 1977 followed in 1978 by 12 Beech T-34C-1 Turbo-Mentors for the basic role. The 24 Alpha Jets will join these types in 1980 to provide advanced training. For the personal use of the King, there is a single Dornier Do28 Skyservant.

Mozambique

Title: Mozambique Air Force
Headquarters: Lourenco Marques
Strength: 35+ aircraft

This ex-Portuguese colony forms one of the main bases for guerrilla operations against neighbouring Rhodesia. With Soviet, Cuban and East European assistance it is steadily building an armed force which includes a fledgling air force equipped initially with a number of ex-Portuguese Air Force aircraft such as Nord Noratlas, Douglas C-47s and Dornier Do27s. Two new air bases have been built, one near the port of Nacala and the other near Beira, and these are expected to accommodate the 35 MiG-21s delivered to the country in 1978. Eight similar aircraft reportedly arrived at Nacala in March 1977, all believed to have come from the Soviet Union.

Nepal

Title: Royal Nepalese Army
Headquarters: Katmandu
Strength: 8 aircraft

The Air Wing of this small, efficient, British-trained army operates two Short Skyvans, one being an Executive version for the Royal Flight delivered in

1970 and the other a standard 3M military version. Three DHC Twin Otters are in service together with an HS748 Srs 2A, the latter received in 1975 and used for passenger/freight duties. Two Aérospatiale SA330 Pumas complete the wing's inventory, one of the machines having a VIP interior for service with the Royal Flight.

Netherlands

Title: Koninklijke Luchtmacht
Headquarters: The Hague
Strength: 330+ aircraft

Combat

The Royal Netherlands Air Force forms part of NATO'S 2nd Allied Tactical Air Force and has two main commands: Tactical Air Command, controlling all combat elements, and Logistic and Training Command. Lockheed F-104G Starfighters predominate in TAC, a total of 120 single-seat fighter-bombers and 18 two-seat TF-104G conversion trainers being procured originally by the KLu, though attrition has reduced the number in service to some 116 aircraft. Two interceptor squadrons, Nos 322 and 323 at Leeuwarden, operate within the Nadge air-defence system of missiles and radars with 18 F-104Gs each plus six reserves. Supplementing these is the USAF's 32nd TFS at Soesterberg, flying 18 McDonnell Douglas F-15 Eagles which replaced a similar number of F-4E Phantoms during 1978. Two nuclear and conventional strike squadrons based at Volkel, Nos 311 and 312 Squadrons, also have F-104Gs. At the same base No 306 Squadron performs reconnaissance duties with 18 RF-104Gs. The Starfighters are due to be replaced by General Dynamics F-16s from 1979, the Netherlands order for this lightweight combat aircraft totalling 80 F-16A single-seaters and 22 two-seat F-16Bs. First deliveries are scheduled for Nos 322/323 Squadrons with a simultaneous change of role from interception to strike.

A total of 75 Canadair-built Northrop NF-5As and 30 NF-5Bs were delivered to the air force between 1969 and 1972 and these aircraft operate in the close-support role in three squadrons: No 314 at Eindhoven, No 315 at Twenthe and No 316 at Gilze Rijen. A fourth unit, No 313 Squadron, acts as an OCU with full combat capability, but operating the two-seat NF-5Bs and based at Twenthe.

Transport and Support

Main transport unit in the KLu's Logistic and Training Command is No 334 Squadron at Soesterberg, equipped with nine Fokker F27M Troopships and three Mk 100 Friendships, although three of the former have been converted into navigation trainers, while the Friendships are used mainly for VIP and government flights. Flown and operated by the air force on behalf of the Army for AOP and liaison are 69 Alouette III helicopters of No 298 Squadron at Soesterberg and No 299 Squadron at Deelen. No 300 Squadron also operates from Deelen, equipped with 30 MBB Bo 105Cs delivered in 1976 and replacing Piper L-21s in the liaison role. KLu aircrew training begins in Canada, with students progressing from Beech Musketeers to Canadair Tutors and CF-5Bs. Operational jet conversion then takes place in Holland using two-seat NF-5Bs and TF-104Gs.

Below: *Northrop NF-5A fighter-bombers of the Royal Netherlands Air Force.* / Koninklijke Luchtmacht

Title: Marine Luchtvaartdienst
Headquarters: The Hague
Strength: 38 aircraft

Above: *Based at De Kooy is this Netherlands Navy Westland Wasp of No 860 Squadron.*
/ Koninklijke Marine

The Royal Netherlands Naval Air Service is a small, compact maritime force of four squadrons equipped with long-range maritime patrol aircraft and helicopters. At the main base at Valkenburg, seven Breguet SP-13H Atlantics of nine originally ordered in 1968, form the equipment of No 321 Squadron. Supplementing these but due for urgent replacement are 15 Lockheed Sp-2H Neptunes of No 320 Squadron. Both types of aircraft have an ASW role and can be armed with AS-12 air-to-surface missiles. Three aircraft are competing for the Dutch Neptune successor, the Lockheed P-3C Orion, Breguet Atlantic Mk 2 and the British Aerospace Nimrod, the first-mentioned appearing the most likely for an order. At Curacao in the Dutch Antilles, there is a detachment of three Neptunes for patrol duties.

Two helicopter squadrons, Nos 7 and 860 at De Kooy, complete the force. The former unit has six Westland UH-14A Lynx, which have replaced seven Agusta-Bell AB204s in the SAR, VIP transport and ship-to-shore liaison roles. A further 10 SH-14B Lynx will enter service in 1979 with uprated engines and the ability to operate in the ASW role, equipped with dunking sonar for use from Navy frigates. Eight more Lynx, of the SH-14C version fitted with MAD equipment, were ordered in January 1978 taking total MLu Lynx orders to 24 machines. Still in use with No 860 Squadron are 10 Westland AH-12A Wasps with detachments assigned to six *Van Speyk* frigates; 12 Lynx are planned to replace the Wasps after 1983. Naval air training is conducted by the national training college for airline pilots on Cessna 150s and Saab Safirs, followed by a course on air force F27s. Helicopter pilots train at Deelen on KLu Alouette IIIs.

New Zealand

Title: Royal New Zealand Air Force
Headquarters: Wellington
Strength: 91 aircraft

Combat

Economic restrictions have prevented any large-scale expansion of the RNZAF and the recent retirement of the long-serving Bristol Freighters and North American Harvards has seen a further depletion in numbers, although the capabilities of the replacement aircraft are greater. The service has a personnel strength of 4,300 and an establishment of seven operational squadrons in the strike, transport and maritime patrol roles under the command of Operations Group at Auckland. The other main element is Support Group, headquartered at Christchurch, which encompasses all units not concerned with front-line operational flying. The present strike element is composed of nine McDonnell Douglas A-4K and four TA-4K Skyhawks flying with No 75 Squadron and based at Ohakea in North Island. Delivered in 1970, the Skyhawks

regularly deploy to Singapore and participate in exercises in Australia. Further north at Whenuapai, Auckland, No 5 Squadron operates in the maritime-reconnaissance role, equipped with five Lockheed P-3B Orions delivered in 1966. When not on military ASW duties, the squadron undertakes mercy missions throughout the South Pacific.

Transport and Support

Whenuapai air station, Auckland, is the main base for No 40 Squadron, which has an establishment of five Lockheed C-130H Hercules and a single communications DH Devon; this unit is assigned the long-range transport task as well as conducting heavy-lift logistical work. At the same base is No 1 Squadron, tasked with trooping and freight missions

Top: *Formation peel off by No 75 Squadron TA-4K Skyhawk, RNZAF.*

Above: *One of two Westland Wasps flown by the Royal New Zealand Navy.*

with six of the 10 HS Andover C1s in use. These aircraft were purchased from the RAF in 1976 and have replaced Douglas C-47s and Bristol Freighters in the short/medium-range role. The remaining four Andovers, plus two Devons, equip No 42 Squadron at Ohakea for a variety of tasks, including VIP and Government work and twin continuation training. At Hobsonville is the RNZAF's main helicopter squadron, No 3, flying 10 Bell UH-1D and UH-1H

Iroquois and six Bell 47Gs. Apart from air force use, the squadron acts as an Army Battlefield Support unit as well as providing a mountain rescue element of two machines at RNZAF Wigram for SAR duties. At Tengah in Singapore is the RNZAF Support Unit with four Bell UH-1Hs providing mobility for the New Zealand Army battalion stationed in the country.

Operational strike training and jet conversion is performed by No 14 Squadron at Ohakea with 16 BAC Strikemaster Mk 88s delivered between 1972 and 1975. Pilot training begins on the 13 CT/4 Airtrainers at the FTS at Wigram while four Bell 47Gs equip the Light Rotary Wing Flight, also at Wigram, for helicopter conversion. Another six Airtrainers have been purchased by the Air Force for spares.

A token Naval air component of two Westland Wasps, delivered in 1966, are deployed aboard the frigates *Canterbury* and *Waikato* for anti-submarine detection and liaison duties. A third is shore-based at Hobsonville and all three come under the wing of No 3 Squadron, RNZAF.

Nicaragua

Title: Fuerza Aerea de Nicaragua
Headquarters: Managua
Strength: 50 aircraft

This small air arm is chiefly committed to civilian duties as part of the National Guard of President Samosa, although it does have a counter-insurgency element which saw action in the 1978 civil war. Almost all the aircraft in service have been supplied by the United States, the most recent addition being a batch, believed to be six, of Cessna 0-2 Skymasters fitted with underwing weapon pylons for the ground-attack role. Six North American B-26 Invaders received some years ago are thought to be grounded. Two dual strike/trainer units have six Lockheed T-33As and six North American T-28Ds, primary tuition being conducted on a few Piper Super Cubs. The transport arm operates one IAI Arava delivered in 1973, four Beech C-45s, three Douglas C-47s and 10 Cessna 180s. Five Casa C212 Aviocars were delivered in 1977-78 to update the force and for Presidential use there is a single HS125 received in 1976. Helicopters include one Hughes 269 for training, four Hughes OH-6As and three Sikorsky CH-34s.

Below: *Nicaraguan Air Force CASA C212 Aviocar.*

Niger Republic

Title: Force Aérienne du Niger
Headquarters: Niamey
Strength: 9 aircraft

Formerly part of French West Africa, Niger continues to maintain close links with France and her air arm is equipped with mainly ex-French AF aircraft. Primarily a transport and liaison arm of the Army, the FAN operates a single Douglas C-54B, three ex-Luftwaffe Nord Noratlas, two Douglas C-47s, an Aero Commander 500 and two Reims F337 Super Skymasters. A single Boeing 737 is used for Government flights.

Nigeria

Title: Federal Nigerian Air Force
Headquarters: Lagos
Strength: 140+ aircraft

Combat

In turbulent Africa, Nigeria stands out as one of the more politically stable countries, the ruling military government persuing a moderate non-aligned stance in the face of continued conflict on the continent. The Nigerian air force combat units operate Soviet aircraft, the principle type being at least 25 MiG-21 interceptors based at Kano and Kaduna and employed in the dual strike/intercept role. Soviet aircraft have been flown by the NAF since the Biafran conflict in the sixties, but it is believed that no more than four Il-28 light bombers and about 12 MiG-17s originally received are now airworthy.

Transport and Support

In September 1975, the Air Force received the first of six Lockheed C-130H Hercules and these conduct heavy-lift operations and supply flights from the main transport base at Lagos. Four Fokker F27 Mk 400s complement the Hercules while a single F28 is flown on VIP duties from the capital. Light transport and internal short-range flights are performed by 20 Dornier Do28 Skyservants, which also operate in the casevac and training roles. For liaison and communications there are 15 Do27s, two Piper Navajos and a single Navajo Chieftain delivered in 1974. The versatility of the helicopter is fully exploited by the FNAF, 10 MBB Bo105Cs ordered in 1973 are employed in the SAR role, 10 SA330H Pumas flying in the troop transport task, while three Westland Whirlwinds still fly on miscellaneous duties. The Bo105D is being assembled locally in Nigeria and 20 aircraft form the intitial batch which will join the 10 C versions in FNAF service.

Primary training is conducted on 32 Scottish Aviation Bulldog 123s, the first 20 machines being delivered in 1973-74 and the remainder in 1978. Advanced flying is done on 10 Piaggio P149Ds, four MiG-15UTIs, about 16 L-29 Delfins and a couple of two-seat MiG-21Us.

Below: *L-29 Delfins of the Nigerian Air Force with red rudders and noses. Behind can be seen five sand and green camouflaged NAF MiG-21MFs.*

Norway

Title: Royal Norwegian Air Force
Headquarters: Oslo
Strength: 209+ aircraft

Combat

A member of NATO's Allied Forces Northern Europe, Norway is divided into two Air Commands, North and South, each fully integrated within the Nadge air defence system. To modernise her combat units, Norway has ordered the General Dynamics F-16 lightweight fighter, and a contract for 60 single-seat F-16As and 12 two-seat F-16Bs was signed in 1977 with deliveries due to begin in 1981. The current RNoAF line-up comprises two fighter-bomber squadrons, No 336 Squadron at Rygge and No 338 Squadron at Orlandet, each with 16 Northrop F-5As, and a third unit, No 718 Squadron at Sola, with 14 two-seat F-5Bs flying in the tactical training role. A total of 108 F-5s were received by Norway but a number have been put into storage. A tactical photo-reconnaissance squadron (No 717) is based at Rygge and is equipped with 13 RF-5As. The air force was also a purchaser of the Starfighter and two units are operational on the type, one assigned the fighter-bomber role and the other having an interceptor and all-weather fighter task. The former unit is No 334 Squadron at Bodo equipped with 22 ex-Canadian CF-104D/Gs acquired in 1972-73, while the latter squadron is 331 Squadron at the same base flying most of the 27 Lockheed F-104Gs and two

TF-104Gs originally supplied to Norway in the 1960s. A maritime patrol unit, No 333 Squadron at Andoya, has five Lockheed P-3B Orions and a requirement exists for three more.

Transport and Support

One transport squadron, No 335 at Gardermoen, operates six Lockheed C-130H Hercules and two calibration and ECM-equipped Dassault Falcon 20Cs. A second unit, No 719 Squadron at Bodo, flies five DHC Twin Otters and eight Bell UH-1Bs on light transport and communications duties. Some 24 UH-1Bs are divided between No 339 Squadron at Bardufoss for Army support and No 720 Squadron at Rygge flying SAR duties. Chief SAR squadron is No 330, flying 10 Westland Sea King Mk 43s delivered in 1972 and based at Bodo, with two aircraft each at Banak, Orlandet and Sola. To update the helicopter force, four Westland Lynx were ordered in 1978 for delivery in 1981. Two more aircraft are on option and all will be of the uprated version for operation from Norwegian Navy ships on coastguard duties.

Basic training is performed on Saab Safirs at Vaernes. AOP and liaison duties are flown by 40 Cessna 0-1E Bird Dogs and Piper L-18C Super Cubs crewed by Army pilots and maintained by Air Force personnel.

Below: *Norwegian Air Force Northrop F-5As of the Jokers aerobatic team.* / B. C. Wheeler

Oman

Title: Sultan of Oman's Air Force
Headquarters: Muscat
Strength: 94 aircraft

Combat

Formed with British assistance in 1958, this air arm has expanded rapidly since 1970 as the country's economy has developed. The 10-year old Dhofar rebellion ended in 1976, and much of the tactical transport capacity used during the war has since been directed into civil aid and development. A small but significant combat element comprises three squadrons equipped with aircraft purchased from Britain over more than 10 years. Backbone of the force are 12 BAC/Breguet Jaguar International strike aircraft made up of 10 single-seaters and two T2 trainers, ordered in 1974, delivered in 1977-78 and presently forming No 8 Squadron at Thumrayt, the SOAF's main strike base. Along with the Jaguars, Oman ordered a £25 million air-defence radar and communications system plus an order for BAC Rapier low-level surface-to-air missiles worth £47 million. Also at Thumrayt are the HS Hunter FGA6 fighter-bombers of No 6 Squadron. A total of 31 Hunters were acquired from Jordan in 1975, but only about 15 are operational at any one time, the remainder being stored. The remaining combat unit is No 1 Squadron at Salalah flying 12 BAC Strikemaster Mk 82As in the ground-attack/advanced training role; a further two are in

Below: *A No 8 Squadron BAC Jaguar International of the Sultan of Oman's Air Force.*

reserve. Oman purchased a total of 24 Strikemasters and apart from attrition, five have been sold to Singapore, delivery being made in 1977.

Transport and Support

Based at Seeb is No 4 Squadron equipped with three BAC One-Eleven 475s delivered in 1974, replacing Viscounts, and operated in the medium and long-range transport role. For short-range work within the country, 15 Short Skyvan 3M freighters equip No 2 Squadron at the same base, while No 5 Squadron operates seven of eight Britten-Norman Defenders delivered late in 1974. Detachments of Skyvans and Defenders are maintained at Salalah which is also the base of the remaining SOAF unit, No 3 (Helicopter) Squadron, flying 20 Agusta-Bell 205s, three AB206 liaison helicopters and five Bell 214A utility machines.

Two Pilatus Turbo-Porters and a single Learjet 25B are in service with the Oman Police Air Wing which has also received recently two DHC-5D Buffalo STOL transports. The Royal Flight has two Swiss AS202 Bravos, two Bell 212s and a single AB212.

Pakistan

Title: Pakistan Air Force
Headquarters: Peshawar
Strength: 360+ aircraft

Combat

Pakistan maintains a non-aligned status with regard to military procurement and the Air Force is in the curious position of operating Chinese, Russian, French and American equipment. Following the 1971 war with India, Pakistan established three air defence sectors in the country, North, Central and South, her present combat units mainly confined to the northern and southern areas while the central sector acts as an early warning area. Most numerous combat aircraft in the PAF is the Chinese-built MiG-19 or Shenyang F-6, which forms the equipment of seven interceptor squadrons. A total of 150 F-6s have been purchased by Pakistan, most being fitted with Martin-Baker ejection seats and American Sidewinder air-to-air missiles. Supplementing these aircraft is the Dassault Mirage element flying in four squadrons assigned strike, intercept and reconnaissance duties. From an initial batch of IIIs purchased in 1969, the present PAF Mirage force is made up of 18 IIIEPs, five IIIDP trainers, 13 IIIRP recce aircraft including 10 delivered in 1977, and 28 5PA strike aircraft. A major Mirage III/5 overhaul facility was officially established at Kamra, near Rawalpindi, in 1978, forming part of the Pakistan Aeronautical Complex and this is expected to handle aircraft operated by other nations as well as those of the PAF.

Based at Masroor is No 7 Squadron, equipped with 15 surviving Martin B-57Bs of 26 originally supplied by America in 1958; they are operated in

Below: *Pakistan Navy Westland Sea King.* / Westland

the night attack role but are badly in need of replacement by a more modern aircraft such as the Jaguar, Mirage F1, etc. Second-line types still in use include four Lockheed RT-33A reconnaissance aircraft based at Peshawar and two squadrons of Canadair Sabres employed in the combat training role at Chaklala air base, near Rawalpindi.

For long-range maritime patrol duties, No 29 Squadron based near Karachi, has three Breguet Atlantics, bought from the French Aéronavale and delivered in 1975. These aircraft operate closely with six Pakistan Navy Westland Sea King Mk 45s, equipped with Exocet anti-shipping missiles.

Transport and Support

Within the PAF transport wing, No 6 Squadron operates at least 14 Lockheed C-130 Hercules, some passed to Pakistan by Iran, and comprising 10 C-130Bs, at least three C-130Es and a single L-100 version. There is a Fokker F27 Mk 200 and a Dassault Falcon 20 for VIP/Government use, while some Beech L-23s and an Aero Commander provide liaison and communications within the Service. Helicopters include ten Kaman HH-43B Huskies for airfield rescue duties, 14 Alouette IIIs plus four in Naval use, 12 Bell 47Gs and a single SA330 Puma for VIP use. In 1976, it was announced that a Hughes 500M helicopter assembly line was to be established in Pakistan and the armed forces expected to receive 50 a year. A similar operation involved the assembly of Cessna T-41D primary trainers for the PAF, 60 a year being planned. Present basic trainer is the Saab Supporter, 45 of which are in service, most at the PAF Academy, Risalpur. Students continue on the Cessna T-37C (30 operated) and Lockheed T-33A (12 operated), with operational conversion flown on four MiG-15UTIs prior to squadron posting.

Title: Pakistan Army Aviation
Headquarters: Dhamial
Strength: 150+ aircraft

Equipped with both fixed-wing and helicopter types, the air element of the Pakistan Army operates in support of ground formations and, in the troop carrying role, has 12 Soviet-supplied Mil Mi-8s acquired in 1969 and the majority of 35 Aérospatiale SA330 Pumas ordered in 1977. Twenty Alouette IIIs assembled in the country plus 20 Bell 47Gs make up the balance of the helicopter force. Fixed-wing liaison, AOP and training are performed by about 50 Cessna 0-1s and some 12 Saab Supporters, the latter being the residue of the PAF total of 45.

Panama

Title: Fuerza Aerea Panamena
Headquarters: Panama City
Strength: 31 aircraft

Formed in 1969 with United States assistance, the FAP has no combat aircraft, its duties being confined to transport, liaison, training and policing. In addition, the air arm is taking over patrols of the Panama Canal as control of it is relinquished by the USA. For long-range transport flights there is a single Lockheed Electra acquired in 1975, while short-range work is performed by four Douglas C-47s, two DHC Twin Otters, two DHC Otters and two Britten-Norman Islanders. Liaison and training is flown on two Cessna U-17Bs and a Cessna 172. Helicopters total 12 Bell UH-1Bs delivered in 1977, two UH-1Ds, two UH-1Hs and one UH-1N Twin-Pac, all of US origin.

Below: *Douglas C-47, one of four in use with the Panama Air Force.* / J. P. Wood

Papua New Guinea

Title: Papua New Guinea Defence Force
Headquarters: Boroko
Strength: 7 aircraft

The air element of the PNGDF comprises one Air Transport squadron, which was formed in August 1975, shortly after the country's independence. The Squadron operates four ex-Royal Australian AF Douglas C-47s and three GAF N22B Nomad utility aircraft. Main roles of this small force are the support of the ground units, training, and land and sea surveillance.

Below: *Australian GAF Mission Master of the Papua New Guinea Defence Force.*

Paraguay

Title: Fuerza Aerea del Paraguaya
Headquarters: Asuncion
Strength: 85 aircraft

Economic difficulties have prevented any great modernisation within the Paraguayan armed forces and the air arm operates a variety of miscellaneous types, emphasis being put on transport, liaison and training aircraft. For counter-insurgency and policing duties, the FAP has a strike/trainer unit flying 12 North American T-6s capable of being fitted with armament but generally used for the training role. For the important transport task there are five ex-

Below: *Paraguayan Navy Cessna 150 trainer.*

Varig Douglas DC-6Bs donated by Brazil in 1976, 10 Douglas C-47s, two C-54s, one Convair 240 bought from Argentina, one DHC Twin Otter and a single Otter.

Helicopters number 14 Bell H-13s and three Hiller 12Es while the training element of the FAP situated at Campo Grande, Asuncion, has eight ex-Brazilian Fokker S-11 primary trainers supplemented by eight Aerotec T-23 Uirapurus purchased in 1975. Advanced tuition is given on 10 T-6s together with the 12 armed versions, and a single MS760 Paris jet. Five Cessna 185s are used for liaison, and the small Paraguayan Navy operates two Bell 47Gs for river patrol plus two Cessna 150Ms for training.

Peru

Title: Fuerza Aerea del Peru
Headquarters: Lima
Strength: 500 aircraft

Combat

The Peruvian Air Force continues to be one of the more modern air arms in South America, purchasing equipment from sources in both the West and the East. Having failed to procure an up-to-date Western combat aircraft in 1976, the FAP ordered Sukhoi Su-22 Fitter C fighter-bombers from Russia, comprising 32 single-seat versions of this variable-geometry aircraft and four two-seat trainers. These are now believed to be flying with two squadrons in Grupo 12, previously an F-86F Sabre unit. To assist in training the Peruvians, 12 ex-Cuban Air Force MiG-21s were delivered to the FAP, accompanied by Cuban technicians. Air defence is vested in two squadrons of Dassault Mirages co-ordinated with a Soviet-installed air defence system of radars and SA-3 missiles. The Mirages began arriving in 1968, and the initial 20 5Ps and two two-seat 5DPs have been increased to a total of 36 aircraft by later re-orders. These fly with Grupo 13 at Chiclayo, and the third squadron in the wing is understood to be flying 18 Cessna A-37Bs. The total Peruvian A-37B order was for 36 aircraft and the remainder equip the third squadron of Grupo 21 at Jorge Chavex air base, Lima. Primary equipment of this wing are two squadrons of BAC Canberras totalling 32

Below: *One of a number of South American countries to fly the Cessna A-37B is Peru.*

Bottom: *Peruvian Navy AB212ASW helicopter; six are in service.* / S. R. P. Thomson

B2/B56/B(I)8s, two T4 trainers and 11 B(I)68s delivered from ex-RAF stocks in 1976-77. Remaining unit is a squadron of 16 HS Hunter F52 fighters purchased in 1956 and forming part of Grupo 12 at Limatambo with the Su-22s. A maritime-reconnaissance unit, Grupo 31, has four Grumman HU-16B Albatross amphibians working in conjunction with the Navy.

Transport and Support

The transport element of the FAP comprises Grupo 41, based at Jorge Chavex air base and equipped with six Lockheed L-100-20s, mainly on pipeline work, delivered in 1973 and 1976-77. A second squadron in the wing has 16 DHC Buffaloes received in the late 1960s, while a third unit operates six Douglas C-47s, a few Curtiss C-46s, five DC-6s and four C-54s. Twenty Antonov An-26 freighters have been supplied by the Soviet Union and these are expected to replace many of the older types in service; a light transport unit (Grupo 42) has seven DHC Twin Otters, some equipped as floatplanes.

AeroPeru, the result of a reorganisation of Satco, the airline element of the air force, has three Fokker-VFW F28s and two F27 Mk 600s for services within the country. Grupo 8 operates 18 Beech Queen Airs on liaison duties, while 12 Pilatus Turbo-Porters are used for light utility work. A Presidential Flight operates a Fokker-VFW F28 delivered in 1976. Helicopters include 17 Bell 212s, 20 Bell 47Gs, 12 Alouette IIIs and six Soviet-supplied Mi-8s, the last mentioned being used mainly for oil-exploration work. Also in service are five Mi-6 heavy-lift machines delivered in 1977.

Training duties are performed by 19 Cessna T-41s, six Pitts S-2As, 26 Cessna T-37Bs, six Beech T-34s, 15 North American T-6s and eight Lockheed T-33As in the armed trainer role. The National Air Photographic Service uses two Gates Learjet 25Bs for survey work.

The Peruvian Army operates in the liaison and communications role and has five Helio Couriers, two on floats, and five Cessna 185s. Eight Bell 47Gs and four Alouette IIIs fly on AOP duties.

Servicio Aeronavale is the airborne element of the Peruvian Navy and received in October 1977, the first of two Fokker F27M Maritime Patrol aircraft ordered the previous year. Nine Grumman S-2E Trackers conduct ASW missions alongside the new machines while from Callao, fleet requirements, support work and training are flown by six Douglas C-47s, a Piper Aztec, and two Beech T-34s. To replace the T-34s, six Beech Turbo-Mentor advanced trainers were delivered in 1978. A further order has been fulfilled for six anti-submarine Agusta-Bell AB212ASW helicopters for use aboard frigates. Other helicopters include 10 Bell JetRangers, two Alouette IIIs, five Bell 47Gs, and six Bell UH-1D/Hs. Four Agusta-built SH-3D ASW helicopters are being acquired with delivery due late in 1978. A total of 42 Mi-8s are in the Naval inventory, of which 36 are currently in store.

Philippines

Title: Philippine Air Force
Headquarters: Pasay City
Strength: 300+ aircraft

Combat

The Philipino armed forces have been recipients of American military aid for more than 30 years and in return, the USA maintains a number of military bases in the country. At present the PAF has seven wings located at five main bases, each wing encompassing as many as five squadrons. The 5th Fighter Wing at Basa Air Base is made up of three Tactical Fighter Squadrons and a training unit: the 6th TFS operates Northrop F-5A fighter-bombers and two-seat F-5B trainers, drawn from respective totals of 19 and

Below: *North American T-28D of the 15th Strike Wing, Philippines Air Force.* / A. Anido

Above: *Philippine Navy B-N Defender.*

three originally delivered from late 1965; the 7th TFS has 20-plus North American F-86F Sabres; the 9th TFS has a counter-insurgency role with nearly 25 converted North American T-34A Mentors; and the 105th Combat Crew Training Squadron has 10 Lockheed T-33As and reconnaissance RT-33As. To modernise the force, 35 ex-US Navy LTV F-8H Crusaders were delivered in 1978 and these are expected to replace some of the older aircraft, initially the Sabres in the 7th TFS.

The 15th Strike Wing has its headquarters at Sangley Point Station and operates two attack squadrons, the 16th and 25th, equipped with about 25 North American T-28Ds. A number of these aircraft, painted all-black for night operations, are based at Zamboanga for Coin missions against the Mindanao National Liberation Front. Also in the wing are the 17th Attack Squadron, flying 16 Siai-Marchetti SF260W Warrior light strike aircraft, and the 27th Search, Rescue and Reconnaissance Squadron, flying four Grumman HU-16B Albatross amphibians. Protection of Philippine airspace is organised under the 580th Aircraft Control and Warning Wing, which is a ground-based unit operating three radar sites at Paredes in the north, Gozar and Paranal.

Transport and Support

There are two transport wings, the 205th and 220th. The former unit is based at Nichols Air Base and operates the following squadrons: the 204th Tactical Airlift Squadron, with seven Fokker F27 Mk 100s; the 206th and 207th Air Transport Squadron, with 30 Douglas C-47s; and the 505th Air Rescue Squadron, with 10 Bell UH-1Hs and some MBB Bo105s. The 220th Heavy Airlift Wing at Mactan Air Base south of Manila is composed of the 221st Heavy Airlift Squadron, equipped with some 15 Fairchild C-123B/K Providers, including some ex-South Vietnamese AF machines; the 222nd HAS, with two Lockheed L-100-20s and three C-130H Hercules; and the 223rd Tactical Airlift Squadron, currently phasing in 12 GAF Nomad Mission Master light utility aircraft. The remaining transport element is the 700th Special Mission Wing stationed at Nichols AB, Pasay City. This formation embraces the 702nd Presidential Airlift Squadron, flying a few Bell UH-1Hs and UH-1Ns, two Sikorsky S-62As, one F27 Mk 200, four NAMC YS-11s, a BAC One-Eleven and a Boeing 707.

Divided between the bases of Nichols and Sangley Point is the 240th Composite Wing, which comprises the 291st Special Air Mission Squadron, with DHC Beavers; the 303rd Air Recce Squadron flying some Douglas AC-47s; the 601st Liaison Squadron with six Cessna U-17A/B Skywagons; and the 901st Weather Squadron, with Cessna 210 Centurions and C-47s. The 100th Training Wing is located at Fernando Air Base, Lipa City. Two squadrons make up the wing: the 101st equipped with about 12 Cessna T-41Ds, and the 102nd Pilot Training Squadron flying 32 SF260MPs. As part of the Philippines Aerospace Development Corp, the Self-Reliance Development Wing of the PAF is planning production of the American Jet Industries Super Pinto jet trainer/Coin aircraft, and an indigenous basic piston-engined trainer design, the XT-100, first announced in 1976.

The Philippine Navy has a small air component equipped with ten Britten-Norman Defenders for SAR and anti-smuggling patrols, and at least three MBB Bo105 helicopters for liaison work.

Poland

Title: Polskie Wojska Lotnicze
Headquarters: Warsaw
Strength: 900+ aircraft

Combat

Poland has the largest air force in the Warsaw Pact, outside the Soviet Union, but like the other countries in the organisation its air arm is principally committed to the tactical role. Supporting the Polish units is the 37th Air Army of the Soviet Frontal Aviation, based at Legnica. The PWL interceptor force consists of some 300 MiG-21s of all marks

equipping nine fighter regiments forming three divisions (approximately three squadrons form a regiment in the Polish Air Force) integrated into an interceptor/missile air-defence system. A few MiG-21RF reconnaissance versions are in service together with about 30 two-seat MiG-21U conversion trainers. Five fighter-bomber regiments have about 200 Sukhoi Su-7s including about 20 two-seat trainer versions, plus a further regiment flying the variable-geometry Su-20 of which there are about 35 in PWL service replacing MiG-17Fs (Polish-built LiM-5s). It is believed that about 60 MiG-17s remain in use as second-line combat trainers.

Transport and Support

Tactical transport duties are performed by one regiment with about 45 aircraft comprising An-12 freighters, An-26 and Il-14 twin-engined transports supported by a few An-2 biplane utility aircraft. A single Tu-134 and about five Yak-40s constitute the VIP/Government flight, while helicopters total some 170 Mi-2s — some flying in the gunship role — Mi-4s and Mi-8s. Short-range liaison work is performed by home-produced PZL Wilgas. Training involves primary flying on WSK TS-8 Bies, basic tuition on the Yak-18, and advanced flying on the WSK-Mielec TS-11 Iskra jet trainer as well as the ubiquitous MiG-15UTI. Light attack duties are flown on single-seat versions of the TS-11.

A shore-based naval air arm is part of the PWL and has one regiment of Polish-built MiG-17s used in the attack role, one interceptor squadron of MiG-21Fs and a few Il-28 bombers for various miscellaneous duties. A number of Mi-2s, Mi-4s and -8s serve in the liaison and SAR roles.

Above: *Single and two-seat Su-7 fighter-bombers of the Polish Air Force.*

Portugal

Title: Forca Aerea Portuguesa
Headquarters: Lisbon
Strength: 270 aircraft

Combat

Portugal's armed forces have suffered from the economic and political difficulties encountered in the country over the past few years. Modernisation has been fragmented and in the case of the FAP, new equipment is urgently needed in certain areas. Front-line air defence is performed by a single unit flying 20 North American F-86F Sabres, based on the west coast at BA-5, Monte Real, of more than 60 delivered from NATO stocks. The FAP is believed to have reverted to squadron numbers in mid-1978, following some years of using base designations. The other jet combat element is situated farther south at BA-6 Montijo, comprising a ground-attack unit (Escuadra 201) flying 18 ex-Luftwaffe Fiat G91R-4s of some 60 originally supplied, and six two-seat G91T-3s (from the same source) used for advanced training and conversion. Also at Montijo is a maritime reconnaissance unit which until the end of 1977 operated four ex-Royal Netherlands Navy SP-2E Neptunes attached to NATO. However, these machines have been withdrawn from service and a replacement is urgently being sought; surplus ex-US Navy Lockheed P-3A Orions are believed to be front-runners, but no decision had been announced by late 1978. Until a decision is made, the five Lockheed C-130H Hercules delivered in 1978, have been tasked with providing SAR and fishery protection. Unsubstantiated reports have also stated that a number of ex-Luftwaffe F-104Gs are likely to be supplied to the FAP, replacing the Sabres on interceptor duties.

Transport and Support

Long-range transport and government flights are flown by the five C-130Hs, first ordered in 1976, planned to replace 10 Douglas DC-6A/Bs currently operated from AB-1, Lisbon. Headquarters of the FAP transport element is at BA-3, Tancos, and types in use include 10 Nord Noratlas of 30 originally supplied from French and German stocks, a few Douglas C-47s and 24 Casa C212 Aviocars ordered in 1974 and currently flying from three bases (BA-1 Sintra, BA-3 Tancos and BA-4 Lajes). A total of 32

Cessna-Reims F337s are in service having been delivered between 1974 and 1976, and now based at BA-2 Ota, BA-5 Monte Real, and BA-7 Sao Jacinto, for liaison and light communications work. Sixteen are equipped to FTG337G Milirole standard for Coin duties and eight operate in the photo-recce role. More than 15 Do27s are flown from BA-3 Tancos and the FAP North American T-6 inventory still stands at some 40 aircraft although not all are believed to be airworthy. Primary training is performed on 30 licence-built DH Chipmunks at BA-2, Ota, while 25 Cessna T-37Cs operate from BA-1, Sintra, in the basic training role. Advanced work is flown on 11 Lockheed T-33As and six ex-USAF Northrop T-38As from BA-5, Monte Real. Helicopter training is performed on about 12 Aérospatiale Alouette IIIs at BA-3, Tancos. Other rotary-wing types include 12 SA330 Pumas flying from BA-4 and BA-6, and 24 Alouette IIIs used for assault and liaison duties from BA-6.

Above: *'Wings of Portugal' Cessna T-37 trainer.* / S. G. Richards

Qatar

Title: Qatar Emiri Air Force
Headquarters: Doha
Strength: 14 aircraft

The QEAF operates a small strike force of two HS Hunter FGA78s and a single two-seat T79 from its main base at Doha on the Persian Gulf. Flown and maintained by British personnel, the Hunters are used mainly on regular coastal patrols. Two Westland Whirlwind IIIs, supplied in 1968, are used for transport duties supplemented by three Westland Commando Mk 2As delivered in 1975. A fourth Mk 2C Commando is used for government and VIP flights. Three Lynx HC28s are flown on utility tasks alongside two SA341 Gazelles operated by the Qatar Police.

Below: *Sand camouflaged Westland Commando on delivery to Qatar.* / Westland

Rhodesia

Title: Rhodesian Air Force
Headquarters: Salisbury
Strength: 120 aircraft

Combat

Anti-guerrilla operations have increased in intensity over the past few years since UDI, not only within Rhodesia itself but also on terrorist camps and installations in neighbouring Mozambique and Zambia. Defence spending has risen accordingly until by 1978, about 26% of the total public spending was being allocated to the armed forces. Long-range bomber-reconnaissance missions and strike duties are the tasks of No 5 Squadron at Old Sarum, Salisbury, with eight BAC Canberras, the survivors of 15 B2s and three T4s originally delivered to Rhodesia in the 1960s. The remaining aircraft have main-spar fatigue problems and are being progressively stripped for spares to keep the others flying. At Thornhill are Nos 1 and 2 Squadrons with nine Hunter FGA9s and eight DH Vampire FB9s respectively; the latter unit also flies eight DH Vampire T55s for OCU-work, although occasionally it is assigned combat duties.

For obvious reasons, great stress is laid on the light attack elements in the RhAF, and these comprise No 4 'Bush' Squadron and No 7 Squadron. The former operates seven Aermacchi AL-60F5 Trojans of 10 originally supplied in 1967 and usually uses them for reconnaissance and FAC duties; a small number of Cessna 182s; and 18 Cessna-Reims F337G Super Skymasters (known as Lynx in Rhodesia) acquired via devious routes in 1976 and fitted out to Milirole standard with underwing hardpoints for the carriage of light stores and rockets. No 7 Squadron is the sole helicopter unit and has taken the brunt of the 'search and destroy' missions, consequently suffering the heaviest loss rate in the air force. Flying Alouette IIIs, of which there are some 34 on strength, the unit co-operates closely with the Army, particularly with elements known as Fireforce. For these counter-insurgency missions the Alouettes act as troop carriers as well as gunships armed with 20mm and .303in machine guns for fire suppression. Fireforce also uses Douglas C-47s of No 3 Squadron for parachuting troops and supplies into difficult areas, and this squadron is believed to have at least eight aircraft on strength.

Transport and Support

As well as the C-47s in use with No 3 Squadron, the unit also has three Britten-Norman Islanders for light transport duties, and a single Beech Baron for VIP use. Basic training is performed on 13 BAC Provost T52s of No 6 Squadron before pupils go to South Africa for conversion on to the Atlas-built Aermacchi MB326 Impala. These Impalas are reportedly owned by Rhodesia but have not been flown outside South Africa and are operated under the auspices of the SAAF.

Below: *Devoid of national markings, a Rhodesian Air Force Hunter on 'stand-by'.* / R. Gardner

Romania

Title: Fortele Aeriene ale Republicii Socialiste Romania
Headquarters: Bucharest
Strength: 550+ aircraft

Combat

A member of the Warsaw Pact, Romania has an air arm which is small compared with those of other states in the bloc. Her modest aircraft manufacturing industry is involved in the development of the IAR-93 Orao light attack aircraft in collaboration with Soko in Yugoslavia (which has design leadership). Since the prototype made its first flight in 1974, development has been slow and the type has yet to enter service; Romania has a requirement for some 80 Oraos to replace MiG-17s currently in service with two fighter-bomber regiments. For interceptor duties, the force operates two regiments with more than 100 MiG-21F/PFs, and in addition to the MiG-17s there is a further regiment with 50 Sukhoi Su-7Bs.

Transport and Support

Two transport squadrons have about 10 An-24s, two An-26s, 30 Il-14s and a single Il-18 VIP transport. The helicopter force has 10 Mi-4s, some Mi-2s and 20 Mi-8s. About 47 Alouette IIIs are armed for the anti-tank role. Training is performed on 60 IAR-823 primary aircraft, 60 Aero L-29 Delfins and 55 MiG-15UTI advanced trainers; operational conversion is flown on two-seat MiG-21Us and Su-7Us. A number of L-200 Moravas are used for communications duties and there is a small Naval SAR unit equipped with Mi-4 helicopters.

Rwanda

Title: Rwanda Air Force
Headquarters: Kigali
Strength: 12 aircraft

This small African state, originally administered by Belgium, gained its independence in 1962. Its modest air arm has undergone limited expansion over the past few years. Three Italian AM3C AOP and liaison aircraft are in service alongside two Douglas C-47s for transport work, two Alouette III helicopters and a Britten-Norman Islander. Three Fouga Magisters were acquired from France in 1975 for jet training, and in 1977 an order for a single B-N Defender was placed.

El Salvador

Title: Fuerza Aerea Salvadorena
Headquarters: San Salvador
Strength: 70+ aircraft

This small Central American republic has had frequent clashes with neighbouring Honduras in the past, and both countries have re-equipped with more modern combat aircraft, despite signing a peace treaty in 1970. The FAS operates a squadron of 18 Dassault Ouragan fighter-bombers purchased from Israel in 1975, and from the same source six Israeli-built Magisters for strike/trainer duties. More recently, three ex-French AF Magisters have been acquired.

Supply and transport duties are performed by a squadron flying a mixed assortment of types including two Douglas DC-6s, 17 C-47s, and four IAI Arava light STOL aircraft. Helicopters include three Aérospatiale Lamas for SAR duties, two Alouette IIIs and a single FH-1100. A training element operates a number of Cessna T-41Cs, 10 North American T-6s and three Beech T-34s.

Saudi Arabia

Title: Royal Saudi Air Force
Headquarters: Riyadh
Strength: 190+ aircraft

Combat

Future front-line combat equipment for the RSAF was assured in May 1978, when the United States Government agreed to the sale of 60 McDonnell Douglas F-15 Eagles to Saudi Arabia to replace the present force of BAC Lightnings. The F-15s are expected to be delivered between 1981 and 1984 and comprise 45 F-15As and 15 two-seat F-15B combat trainers. As well as purchasing military equipment for herself, Saudi Arabia finances other Arabian countries including Egypt, and was

Below: *Royal Saudi Air Force BAC Strikemaster Mk 80As.* / BAC

instrumental in procuring Mirages and Sea King helicopters for the Egyptian Air Force.

The main interceptor force in the RSAF is composed of BAC Lightnings and Northrop F-5s. A total of 31 Lightning F53s (32 delivered), two F54s and five two-seat T55s are in service, of which 20 F53s are operational with No 2 squadron at Tabuk in the north on interception duties. The remaining aircraft operate from Dharan at the OCU. Based at Taif and Khamis Mushayt are two fighter-bomber squadrons Nos 3 and 10, with a further two, equipped with 70 Northrop F-5Es, being formed. These aircraft are equipped to carry Maverick guided bombs under a contract with Northrop, which also services and trains Saudi nationals on the type. A similar agreement exists between BAC and the Saudi Government over the Lightning/Strikemaster force.

Transport and Support

Two transport squadrons, Nos 4 and 16, are based at Jeddah and equipped with nine Lockheed C-130Es. 25 C-130Hs and four KC-130H Hercules, the latter for F-5 air refuelling. Based at Taif are Nos 12 and 14 Squadron, flying 16 Agusta-Bell AB206 JetRangers used for helicopter training and 24 AB205s for liaison, SAR and airfield crash rescue duties. A number of AB212s have been acquired for coastal SAR work and six Kawasaki KV-107s, plus a further six planned, have been ordered for similar duties. Saudi Arabia will also benefit from the Westland Lynx production order announced in 1978 in which Egypt will manufacture this helicopter during the 1980s for itself, Saudi Arabia and other Gulf States. The Royal Flight, alias No 1 Squadron, has a single Boeing 707-320, two Lockheed JetStars, an AB206 JetRanger, two AS-61A-4s delivered in 1977 and, due shortly, a specially equipped Boeing 747SP.

Two training units with a secondary strike potential are Nos 9 and 11 Squadrons at Riyadh, which forms part of the King Faisal Air Academy. A total of 46 BAC Strikemaster Mk 80As have been purchased by the RSAF; No 9 Squadron acting as a basic jet training unit, while No 11 Squadron operates in the weapon training role. Also at Riyadh is No 8 Squadron which is part of the KFAA and designated a primary training unit with an establishment of 13 Reims-built Cessna 172G/H/Ms. Dharan-based No 7 Squadron, with 24 Northrop F-5F two-seat conversion trainers, acts as the F-5 OCU, and its companion unit, No 15 Squadron, flies 20 older F-5B trainers, four of which may be transferred to the Yemeni Air Force.

Senegal

Title: Armée de l'Air du Senegal
Headquarters: Dakar
Strength: 23 aircraft

As with many ex-French colonies, Senegal has received a number of ex-Armée de l'Air aircraft and still benefits from a training agreement with France. The country's small air arm has a personnel strength of some 200 and is mainly concerned with internal policing, transport and liaison work. Two ex-French Magister jet trainers are in use alongside six Douglas C-47s, four MH1521M Broussards, two Alouette II helicopters, one SA342 Gazelle and a Reims F337 Super Skymaster counter-insurgency aircraft. A single Boeing 707-200 is used for government and VIP flights and is based at Dakar. Towards the end of 1977, six Fokker-VFW F27 Friendships began delivery to the air force, order completion planned for early 1979.

Sierra Leone

Title: Sierra Leone Defence Force
Headquarters: Freetown
Strength: 7 aircraft

Formed in 1973, this small air arm operates four Saab MFI-15 primary trainers and one Hughes 300 and two Hughes 500 helicopters.

Singapore

Title: Republic of Singapore Air Force
Headquarters: Singapore
Strength: 155 aircraft

Combat

This small island nation officially established an Air Defence Command in September 1971 and subsequently renamed the arm the RSAF. The current strike force comprises two squadrons of HS Hunter Mk 74s based at the main strike airfield of Tengah. These are No 140 (Osprey) Squadron, which also has four FR74As on strength, and No 141 (Merlin) Squadron. About 31 single-seat Hunters are in service, plus seven two-seat T75s and the four reconnaissance FR74As, of a total procurement of 47 aircraft. At the same base are Nos 142 and 143 Squadrons equipped with 40 McDonnell Douglas A-4S Skyhawk fighter-bombers and five of six TA-4S two-seat combat trainers delivered in 1974-76. These aircraft have been converted from ex-US Navy A-4Bs by Lockheed Aircraft Service Singapore (LASS), incorporating more than 100 modifications including a new avionics fit, 30mm Aden cannon and the installation of uprated 8,400lb J65 engines. Planned for delivery early in 1979 are 21 Northrop F-5s ordered in 1976 at a cost of $113 million. The 18

F-5Es and three two-seat F-5F trainers will replace one of the Hunter squadrons and operate in the interceptor role for which they will be equipped with AIM-9J Sidewinder missiles, 200 of which are included in the deal.

Above: *Singapore Air Force C-130B Hercules in desert colours which betray its previous Jordanian owners.* / H. Holmes

Transport and Support

Transport duties are performed by 121 Squadron which received in 1978 four Lockheed C-130B Hercules, two from ex-United States Air Force stocks and two Bs from Jordan. These have been acquired for Singapore's ANZUS commitment. Six Short Skyvan 3Ms also serve in the unit at Changi being used for SAR and anti-smuggling duties around the island. The RSAF's helicopter force is encompassed within No 120 (Condor) Squadron, also at Changi, and operates seven Alouette IIIs, three Bell 212s and at least 12 Bell UH-1Hs. The unit supplies machines for planeguard duties at the main bases, the Bell 212s having a VIP transport role.

At Seletar, the main training base, No 130 (Eagle) Squadron has about 20 BAC Strikemaster Mk 84s for basic jet training and weapon firing, of a total of 25 aircraft purchased by the RSAF of which four came from South Yemen and five from Oman. Flying training begins on 14 Siai-Marchetti SF260MS piston aircraft based at Changi and forming No 150 (Falcon) Squadron.

Somalia

Title: Somalian Aeronautical Corps
Headquarters: Mogadishu
Strength: 112+ aircraft

Although appealing for more modern equipment during and after the 1977 war with Ethiopia, Somalia had not significantly changed its front-line air force inventory up to late 1978. The combat elements continue to fly the Soviet equipment supplied prior to the war including an interceptor squadron with 10 MiG-21s, two squadrons of MiG-17s and MiG-19s, and a bomber squadron with 10 Il-28s flying from Mogadishu. Serviceability is believed to be low following the cut-off of spares from Russia late in 1977. Transports include three Douglas C-47s, a Beech C-45, at least three An-24s and An-26s, and three An-2s. Eight Piaggio P148s are used for primary training, together with at least one Cessna 150 Aerobat, while for advanced flying there are 20 Yak-11s and a couple of MiG-15UTIs. Helicopters include Mi-4s, Mi-8s and some Agusta-Bell AB204s. Two Dornier Skyservants are flown by the Police Air Wing for liaison and supply duties.

South Africa

Title: South African Air Force
Headquarters: Pretoria
Strength: 650+ aircraft

Combat

The ban on arms exports to South Africa was fully implemented by the United Nations in 1978, but the country is almost completely self-sufficient in arms production, particularly with regard to aircraft. The Atlas Aircraft Corp at Johannesburg is currently building the single-seat Aermacchi MB326K or Impala II against an initial SAAF order for 50 aircraft, production of the two-seat Impala I having been completed with 151 built. Also being produced is the C4M Kudu light transport, based on the Aermacchi AL60 with the AM3C wing and tailplane. Even more important is the assembly of Dassault Mirage F1AZ strike aircraft, with eventual manufacture for a SAAF requirement that is expected to total some 100 aircraft.

Strike Command, comprising five squadrons, has its HQ at Waterkloof, which is also the SAAF's main

Mirage F1 Base. No 1 Squadron is steadily re-equipping with 32 F1AZs as they come off the Atlas assembly line and the unit shares the airfield and support facilities with No 3 Squadron which has 16 F1CZ interceptors. No 2 Squadron at Waterkloof fulfils both fighter-bomber and reconnaissance roles with a Mirage III complement of 16 IIICZs, four IIIRZs, four IIIR2Zs and three IIIBZ trainers. The remaining Mirage unit is 85 Advanced Flying School at Pietersberg which has 16 Mirage IIIEZs, 10 IIID2Zs and three IIIDZ trainers, making a total of 56 of these French aircraft in service. Other Strike Command units are Nos 12 and 24 Squadrons at Waterkloof, operating six BAC Canberra B(I)12 bombers and three T4 trainers flying with the former squadron, and eight HS Buccaneer S50 strike aircraft with the latter unit. To increase the effectiveness of the Command's units, a new airbase has been built at Hoedspruit in the north-east of the country.

Maritime Command operates from D. F. Malan Airport, Cape Town, with No 35 Squadron flying long-range MR duties with seven HS Shackleton MR3s, recently re-sparred to extend their useful lives for a few more years. At Ysterplaat Air Station, Cape Town, No 27 Squadron operates 20 Piaggio P166S Albatross for in-shore reconnaissance and liaison duties, with No 25 Squadron flying six Douglas C-47s in the fleet requirements rôle. Eleven Westland Wasp HAS1s equip 22 Flight, serving aboard South African Navy ships in the ASW and communications role, and shore-based at Ysterplaat.

Transport and Support

Air Transport Command HQ is at Waterkloof controlling three squadrons, Nos 21, 28 and 44. At the Command's main base and HQ is No 28 Squadron flying seven Lockheed C-130B Hercules and nine C160Z Transalls for tactical and long-range work, while at Zwartkop No 44 Squadron operates five Douglas DC-4s and ten C-47s. At the same base is No 21 Squadron which undertakes VIP and government transport flights using one BAC Viscount 781, four HS125s and five Swearingen Merlin IVAs. The latter were delivered during 1975-76, and one is fitted out as an ambulance aircraft. Two further Merlins are civilian-registered for Government use.

The helicopter element of the SAAF flies mostly French types, the largest being the Aérospatiale Super Frelon, 14 of which conduct heavy-lift and casevac duties with No 15 Squadron at Zwartkop and Bloemspruit. Tactical troop transport is flown by 20 SA330 Pumas of No 19 Squadron divided into two flights at Zwartkop and Durban. A further 20 Pumas have reportedly been received to double the inventory of this type. There are two squadrons of Alouette IIIs: No 16 Squadron with 20 aircraft at Ysterplaat and Port Elizabeth, and No 17 Squadron with 10 aircraft at Zwartkop. More Alouettes are believed to be in service than the number given.

Light Aircraft Command has its headquarters at Zwartkop controlling No 41 Squadron at Lanseria with 20 AM3C Bosboks and some Kudus, No 42 Squadron at Potchefstroom with 20 Bosboks, No 43 Squadron at Durban with Cessna 185s and No 11 Squadron with 20 Cessna 185s at Potchefstroom. A total of 40 Kudus are on order from Atlas.

Training Command has nearly 100 North American Harvards at the Flying Training School at Dunnottar, and 80 Atlas Impala I/IIs at the FTS at Langebaanweg. There are three Advanced Flying Schools: No 85 AFS at Pietersberg with 12 Sabre 6s and the Mirages; No 86 AFS at Bloemspruit with six C-47s for twin conversion; and No 87 AFS at the same base with 10 Alouette IIIs for helicopter training. The Active Citizen Force has almost changed over from Harvards to Impalas. The following units each have 15 aircraft: No 4 Squadron at Lanseria, No 5 Squadron at Durban, No 6 Squadron at Port Elizabeth, No 8 Squadron at Bloemspruit and No 7 Squadron at Ysterplaat. The only remaining Harvard unit is No 40 Squadron at Dunnottar, with 10 aircraft. There are 13 Air Commando Squadrons, Nos 101-112 plus No 114 (Women's) ACS, under the Command of the SAAF and equipped with civil light aircraft for use in emergencies.

Below: *No 3 Squadron Dassault Mirage F1CZ based at Waterkloof, South Africa.*

Soviet Union

Title: Soviet Military Aviation Forces
Headquarters: Moscow
Strength: 15,000+ aircraft

Combat

Numerically the largest air arm in the world, the Soviet Air Force continues to expand at an alarming rate, the new breed of combat aircraft replacing older types at a greater ratio than one-for-one. It is estimated that Russia is manufacturing more than 1,000 new machines annually, the majority being of an offensive rather than defensive nature. More advanced aircraft are under development for service in the late 1980s, including a smaller Foxbat-like ground-attack aircraft with fighter potential, a pure tank-buster along similar lines to the Fairchild A-10, and a new bomber of canard configuration.

The Soviet Air Force is made up of five separate major elements each having a different role, command structure, and equipment: Long-Range Aviation (Aviatsiya Dal'nevo Deistviya), Air Force of the Anti-Aircraft Defence of the Homeland (Istrebitel'naya Aviatsiya P-VO Strany), Frontal Aviation (Frontovaya Aviatsiya), Air Transport Aviation (Voennotransportnaya Aviatsiya) and Naval Aviation (Aviatsiya Voennomorskovo Flota). The unit structure is based on divisions each comprising three or more regiments, which in turn consist of three squadrons each with about 12 aircraft.

Long-Range Aviation is divided into three groups, two in western Russia and one in the east, operating about 700 aircraft. Most modern type in service is the variable-geometry Tupolev Tu-26 Backfire-B bomber. Although the Soviet Union has insisted that the Backfire is a tactical machine, western intelligence sources have indicated that the aircraft has an unrefuelled radius of action of some 3,750 miles and is capable of carrying stand-off missiles, ending any doubt as to its strategic potential. The aircraft is believed to equip at least one Air Force regiment with about 50 aircraft. Production is believed to be running at between two and three aircraft a month. In addition there are about 110 Tupolev Tu-95 Bears and 35 Mya-4 Bison-Bs and -Cs in the bombing role, plus 45 Bison-As and a number of Tupolev Tu-16 Badgers employed in the flight refuelling role. Approximately 140 Tu-22 Blinders and less than 300 older Tu-16s operate in the medium-range bomber role. These types are scheduled to be replaced by Backfires, possibly on a one-for-one basis. A tanker version of the Ilyushin Il-76 Candid transport is being developed to replace the Bisons.

Frontal Aviation is the Soviet Union's tactical air force and, with 5,500 aircraft plus a further 3,000 second-line types, forms the bulk of Russia's aviation arm. Close support, interdiction and tactical strike are the main roles of the FA, and this force is currently undergoing a major re-equipment programme with new types replacing older aircraft. One of the most significant is the v-g Sukhoi Su-19 Fencer fighter-bomber, of which more than 250 are in use. Another variable-geometry aircraft is the MiG-23 Flogger, of which there are two main versions in FA service: the single-seat -23S Flogger-B fighter and the two-seat -23U Flogger-C trainer. A single-seat ground-attack version is known as Flogger-D and designated MiG-27; Flogger deliveries to the SovAF total about 1,000+ machines. It is expected that this type will eventually replace the 1,500-plus MiG-21s now in service. About 400 Su-7s remain in use but their gradual withdrawal is underway, their place being taken by the v-g Su-17 Fitter-C. Other types include MiG-25 Foxbat-B reconnaissance aircraft, and about 170 Yak-28 Brewer-D and -E reconnaissance and ECM aircraft. Two regiments of Mi-24 Hind helicopters are operational in East Germany, supported by a few hundred Mi-8 Hip assault machines; there are some armed Mi-24s known as Hind-D for attack duties.

Air Defence Forces of the Homeland, as well as deploying numerous surface-to-air missiles, operate an interceptor force of some 3,300 aircraft, comprising about 1,000 Sukhoi Su-15 Flagon-As, -Ds and Es, 1,500 MiG-23/25s, some 150 Tu-28P Fiddler long-range interceptors, and some 700 Su-9/Su-11 Fishpots. There are also some Yak-28s, a few Tu-22s and about 12 Tu-126 Moss airborne early-warning aircraft.

Naval Aviation has more than 1,250 aircraft, mostly based near the north-west and Black Sea coasts and organised into three regiments each with three squadrons at each base. The aircraft support four Soviet Fleets — the Baltic Red Banner Fleet, Northern Fleet, Black Sea Fleet and Far East Fleet. For long-range reconnaissance, approximately 45 Tu-95 Bear-Ds and 15 Bear-Fs are operated with 150 Tu-16 Badger-C, -D, -E and Fs, some of which are flown as tanker aircraft. About 275 Tu-16 Badger-Gs are used in the strike role equipped with one Kingfish, one Kipper or two Kelt air-to-surface missiles, while nearly 50 Tu-22s fly on strike-reconnaissance, some with Kitchen ASMs. A few Mya-4 Bison-B/Cs fly long-range surveillance missions and under naval command there are a few Tu-126 Moss and about 50 Backfire bombers. Recently transferred to Naval Aviation were some v-g Sukhoi Su-17 Fitter-C fighter-bombers for the close-support of Marine units. Maritime reconnaissance duties are conducted by 60 Il-38 Mays and about 100 Beriev Be-12 Mail amphibians. ASW helicopters include Mi-14 Haze, Ka-25 Hormone and Mi-4 Hound modified for the role.

Transport and Support

Military Transport Aviation has about 1,700 aircraft. There are 800 An-12 Cubs, plus a small number of An-26s, An-24s, An-14s, Il-14s and Il-18s. About 30 An-22 Cocks are in service for heavy-lift duties; the new Il-76 four-jet transport, which joined the ATA in 1975, continues to join the force in ever-increasing numbers and is expected to supplement and eventually replace the An-12s. A new light tactical jet transport, designated An-72, was revealed in 1977 and is expected to join the SovAF to replace the An-24/26 series. To complement the fixed-wing force, a helicopter element has some 2,500 machines, including 600 used by Frontal

Above: *Soviet Northern Fleet Tu-16 Badger.* / Tass

Aviation. Reliance is placed in the main on the Mi-8 but large numbers of Mi-4s are still in service for a variety of tasks. Other types include about 500 Mi-1/2s, some Mi-10/12 heavy-lift machines, and Mi-6s and -24s. Numerous liaison and communications aircraft operate throughout Russia. Training is conducted on Yak-18 Max for primary flying, L-29 Mayas for basic work (being replaced by the new L-39) and MiG-15UTI Midgets for advanced conversion. OCU flying is done on two-seat versions of the main combat aircraft.

Naval Aviation has a number of communications and support aircraft totalling 200 while about 250 helicopters are in use.

Title: Soviet Navy
Headquarters: Moscow
Strength: 500+ aircraft

Four 60,000 ton aircraft carriers known as the Kuril class are due for naval service, with the first vessel, named *Kiev*, forming part of the Black Sea Fleet for deployment in the Mediterranean. A second, named *Minsk*, is now on trials (1978) and the *Kharkov* is due to be launched imminently. The fourth much larger ship, with more accommodation for aircraft, is under construction. *Kiev* carries about a dozen fixed-wing Vtol combat aircraft attributed to Yakovlev and code-named Forger-A; a two-seat trainer version, Forger-B, is also in use and total production of the type is believed to be no more than about 30 machines. Supporting these are Kamov Ka-25 Hormone ASW helicopters which also form the complement of the two anti-submarine helicopter cruisers, *Moskva* and *Leningrad*, each capable of carrying up to 20 Ka-25s.

Spain

Title: Ejercito del Aire
Headquarters: Madrid
Strength: 750+ aircraft

Combat

Spain is divided into three air regions; No 1 headquartered in Madrid, No 2 in Seville and No 3 in Zaragoza, with a fourth region covering the Canaries. Within these regions Air Defence Command operates an interceptor force in conjunction with the national air-defence system, known as Combat Grande. Purchased through the US Air Force, this system consists of seven long-range radar sites supplying data round-the-clock to a central computer, from where the information is transmitted to the appropriate Air Sector. The interceptor force is made up of two squadrons (known as Escuadrones), Nos 111 and 112 Squadrons of the 11th Wing (known as Ala) at Manises, flying 22 Dassault Mirage IIIEEs and six IIIDE two-seat trainers, and two squadrons, Nos 121 and 122, forming No 12 Wing at Torrejon with 33 McDonnell Douglas F-4CR(S) Phantoms of 36 delivered. At Los Llanos in the south is No 14 Wing with No 141 Squadron, equipped with 14 Dassault Mirage F1CEs, with No 142 Squadron forming with a further nine aircraft ordered in January 1977; 48 more Mirages are on order. An operational conversion group, Grupo 41, is attached to ADC at Zaragoza and equipped with more than 20 Lockheed T-33As. To supplement the Phantoms, a

Above: *Spanish Air Force F-4C(S) Phantom of No 121 Squadron, Torrejon.* / Ejercito del Aire

further four F-4Cs and four RF-4Cs are being delivered, but Spain refused America's offer of 42 F-4Es.

Tactical Command comprises No 21 Wing based at Moron, with No 211 Squadron flying nine Northrop SF-5As, nine SRF-5As and SF-5Bs, and No 214 Squadron with 23 Hispano HA220 Super Saetas. Based in the Canary Islands at Gando for air support and reconnaissance is No 46 Wing, which is made up of four squadrons: No 461 Squadron with 12 Casa Aviocars, No 462 Squadron with 15 HA200D Saetas, No 463 Squadron with 20 North American T-6Ds, and No 464 Squadron with nine SRF-5As, eight SF-5As and two SF-5Bs flying in the tactical recce. role. At La Parra No 221 Squadron with nine Grumman HU-16Bs and two of three ex-US Navy Lockheed P-3A Orions delivered, flies on ASW duties manned by joint EdA/Navy crews. Also attached to Tactical Command is a liaison flight at Tablada (Flight 407) with seven Cessna O-1Es and 12 Do27s.

Transport and Support

Air Transport Command encompasses two wings, Nos 35 at Getafe and No 37 at Villanubla. Two units make up No 35 Wing, No 351 Squadron with nine Casa C212 Aviocars, and No 352 Squadron with a similar number of Aviocars. Total EdA procurement of this light transport stands at 71. At Villanubla is No 37 Wing with No 372 Squadron flying 12 DHC Caribous and No 371 Squadron with Aviocars. Autonomous within the Command is No 301 Squadron, the Lockheed C-130H Hercules unit based at Zaragoza, which is equipped with four transport versions and three KC-130H tankers for flight refuelling of the Phantoms. No 91 Group at Getafe has No 911 Squadron with four Aviocars and eight Casa Azors, and No 912 Squadron with five Piper Aztec Es, a Piper Navajo, 61 T-6s and five Beech Barons.

Within Training Command is No 791 Squadron at San Javier, whose task is basic training using 29 Beech F33C Bonanzas and 25 Beech T-34As; No 792 Squadron has a dual role, eight AISA I115s operating in the basic role and five Aviocars flying as crew trainers; No 793 Squadron has 37 T-6Gs, 30 T-6Ds and 62 HA200s for conversion training. Jet conversion is done on 23 SF-5Bs of Nos 731 and 732 Squadron at Talavera la Real, while multi-engine training is performed by No 741 Squadron with eight Aviocars and No 742 Squadron with 24 F33A Bonanzas, eight Beech King Air C90 instrument trainers and 18 Beech Baron trainers. A total of 60 Casa C-101 Aviojet basic/advanced trainers are on order, with the first unit becoming operational in 1980. At Cuatro Vientos Nos 751 and 752 Squadrons have 24 AB47Gs and three Bell UH-1Hs for helicopter training, the former type being replaced by 17 Hughes 300s ordered in 1978.

Servicio de Busqueda y Salvamento or Search and Rescue Service operates 14 AB205 helciopters on ASR and VIP duties alongside nine SA-16A Albatross, five Alouette IIIs, three AB47J and two Do27s flying in three units (Nos 801, 802 and 803 Squadron) from Mallorca, the Canary Islands and Madrid respectively. Three Fokker-VFW F27M Maritime SAR aircraft are due for delivery in 1979, replacing the Albatross with one each at the above bases. In addition there are 11 other units, tasked mainly with support and liaison duties and including No 403 Squadron at Cuatro Vientos with five Do27s, five Aviocars for photographic work, No 404 Squadron at Torrejon with seven Canadair CL-215 amphibians for fire-fighting and tactical duties, and No 721 Squadron at Alcanterilla with Aviocars for paratroop training. For VIP flights one Douglas DC-8 and four Dassault Falcon 20s are on the strength of No 401 Squadron.

Title: Arma de la Armada
Headquarters: Madrid
Strength: 61 aircraft

Spain's Naval Air Arm operates the 15,890 ton helicopter carrier *Dedalo* (ex-USS *Cabot*), transferred in August 1967, which, with the acquisition of V/STOL Matadors, has been reclassified as an aircraft carrier. The six Hawker Siddeley AV-8A Matadors (reduced to five following a crash in 1976) and two TAV-8A two-seat combat trainers, ordered via the United States in 1973, form No 008 Squadron. A further five Matadors have been ordered and eventual deliveries could take the total to 24 aircraft. For ASW and SAR duties there are 12 Sikorsky SH-3D Sea King helicopters, shore-based at Rota naval base and designated No 005 Squadron, with detachments embarked on *Dedalo*. No 001 Squadron has 12 Bell/AB47s for pilot training and communications duties, while No 003 Squadron

Above: *McDonnell Douglas AV-8A Matador of the Spanish Navy.*

operates three of four AB212ASW anti-submarine helicopters delivered in 1974 plus four AB204B transport machines for Marine assault use. A further 12 AB212ASWs have been ordered to increase the effectiveness of the force. Another ASW unit is No 006 Squadron, flying 13 Hughes 500M(ASW) light torpedo-armed helicopters received in 1972, and deployed aboard Spanish Navy destroyers. For attack duties six Bell AH-1G HueyCobras of eight delivered form No 007 Squadron. Fixed-wing liaison is performed by No 004 Squadron, with two Piper Comanches and two Twin Comanches.

Title: Fuerzas Aeromoviles del Ejercito de Tierra (FAMET)
Headquarters: Los Remedios
Strength: 113 aircraft

Established in July 1965, this aviation branch of the Spanish Army is a helicopter force with headquarters near Madrid and made up of five Helicopter Units or Unidades de Helicopteros. At Los Remedios is Unidad I, equipped with three Bell OH-58As, 16 Bell/AB205s and two Alouette IIIs for a variety of duties including close-support and liaison. Unidad II is at Virgen del Camino, with three Bell OH-58As and 12 Bell UH-1Hs; a similar establishment is flown by Unidad III at Agoncillo and Unidad IV at El Copero. FAMET's heavy transport flight is Unidad V at Los Remedios, equipped with ten Boeing CH-47C Chinooks. A breakdown of FAMET helicopter strength in addition to the Chinooks is 52 Bell UH-1Hs, six UH-1Bs (used for training), 12 Bell OH-58A/AB206s, six Bell 47Gs and two Alouette IIIs. A further 18 OH-58As and eight UH-1Hs are being delivered.

Sri Lanka

Title: Sri Lanka Air Force
Headquarters: Colombo
Strength: 62 aircraft

Five Soviet MiG-17F fighter-bombers supplied in 1971 form the equipment of Sri Lanka's only combat unit. A jet trainer squadron operates eight BAC Jet Provost T51s and a two-seat MiG-15UTI. The transport force, which undertakes tourist flying in addition to its military tasks, is equipped with a Convair 440, five DH Doves, two DH Herons, two Riley Herons, four Cessna Super Skymasters and two ex-Air Ceylon DC-3s. The helicopter fleet has been expanded with the purchase of two Aérospatiale SA365 Dauphin IIs from France, and these join two Kamov Ka-26s, seven Bell JetRangers and six Bell 47Gs. Training is performed on nine DH Chipmunks and six Cessna 150s.

Below: *Flown on both civil and military duties, a Sri Lanka Air Force Convair 440.*

Sudan

Title: Sudanese Air Force
Headquarters: Khartoum
Strength: 100+ aircraft

Combat

A major policy shift by the Sudan occurred during 1977-78 when reliance on Russian arms supplies ceased and the country moved firmly into the Western sphere of influence. Saudi Arabia is helping to finance the modernisation of the Air Force and orders have been placed in France for 14 Dassault Mirage 50 fighter-bombers plus a further 14 on option, together with ten Aérospatiale SA330 Puma assault helicopters. The United States has also stated its willingness to sell the Sudan a squadron of Northrop F-5Es, but no order has yet been officially announced. The present SAF inventory comprises two combat squadrons, one flying about 15 MiG-21PF interceptors, and a fighter-bomber unit equipped with 12 Chinese-built MiG-17Fs (Shenyang F-4s). Of five BAC 145 counter-insurgency aircraft supplied in the late 1960s, only three remain in flyable condition.

Transport and Support

Ordered in April 1977, were six Lockheed C-130H Hercules which began arriving in March the following year, supplemented by four DHC-5D Buffaloes, all allocated to the SAF transport squadron. This unit had received six An-12s, five An-24s and ten Mi-8 helicopters from Russia but serviceability had deteriorated after the Soviet withdrawal. Other types in use include a single DHC Twin Otter 300 survey machine, eight Pilatus Turbo-Porters for light transport duties and a Beech King Air 90 flown by the Police Air Wing. Twenty Bo105 helicopters are on order. The eight BAC Jet Provost T52s are believed to be in storage.

Below: *The first DHC-5D for the Sudanese Air Force.* / T. Wildman

Sweden

Title: Flygvapen
Headquarters: Stockholm
Strength: 740+ aircraft

Combat

This modern, efficient air arm is supported by a strong Swedish aircraft industry which has produced all three main combat types currently in service: the Saab Viggen, Draken and Saab 105, totalling just under 600 aircraft. Flygvapen or Swedish Air Force organisation is based on Flottiljer or Wings with two or three squadrons of about 18 aircraft each. Future defence plans announced in 1977 stated that the number of medium attack squadrons would remain at six, though the 17 fighter units at present in service will be reduced to 10 by 1982 and to nine by 1987. The reconnaissance units are also to be reduced, from eight to six by 1982.

Saab 37 Viggens are steadily replacing the Draken fleet, the main versions being the AJ37 all-weather attack aircraft, replacing Lansens; SK37 two-seat trainer; SF37 reconnaissance aircraft; SH37 sea-surveillance and attack aircraft; and the JA37 interceptor planned for eight squadrons with deliveries under way. More than 100 of 180 SK/AJ/SF/SH Viggens on order are now in service; 149 JA interceptors are on order with deliveries beginning in 1978.

For attack and reconnaissance duties the SAF has the following wings: F6 with two squadrons of AJ37

Viggens at Karlsborg; F7 with two squadrons of AJ37s at Satenas; F11 assigned to the reconnaissance role with one squadron of S32C Lansens and one squadron of S35E Drakens at Nyköping; and F15 with one attack squadron of AJ37s and the Viggen conversion unit flying SK37s at Soderhamn.

For air defence 17 interceptor squadrons equipped with about 300 Saab J35D/F Drakens are deployed throughout the country within six regional military districts incorporated into the Stril 60 air-defence system. They are: F1 at Vasteras, with two J35F squadrons; F4 at Ostersund, with three J35D squadrons; F10 at Angelholm, with three J35F squadrons; F12 at Kalmar, with two J35F squadrons (due for disbandment in 1978); F13 at Norrköping, with two J35F squadrons and a reconnaissance unit with SH/SF37s; F16 at Uppsala, with two J35/SK35 squadrons, including the Draken OCU; F17 at Ronneby, with two J35F squadons and an SH37 squadron; and F21 at Lulea, with a recce squadron of S35Es, a fighter squadron flying J35Ds and a light attack squadron with Sk60Cs (Saab 105s). As well as the single units F13 and F17, SH37s will also equip one squadron in F21.

Transport and Support

Attached to F7 at Satenas, is a transport squadron flying two Lockheed C-130E and one C-130H Hercules, and seven Douglas C-47s. Another unit flies two ex-SAS Caravelles on high-speed transport and VIP flights from Malmslatt, designated F13M

Above: *Saab SF37 Viggen reconnaissance aircraft of the Swedish Air Force.* / Saab-Scania

Wing. The same unit has a squadron of 24 target-towing J32B Lansens, including about 12 for ECM work, and a few Sk60Cs. Training in the Swedish Air Force is undertaken at the basic stage by F5 at Ljungbyhed FTS equipped with Scottish Aviation Sk61 Bulldogs, followed by advanced tuition on Saab Sk60s of the same unit. The Air Force College at Uppsala has Sk60s for training with F20, while F18 at Stockholm have a number of the same aircraft type for liaison duties together with a few Saab Safirs. A total of 58 Bulldogs were delivered to the Service, and there are almost 150 Saab Sk60 trainers and 20 attack versions.

The Royal Swedish Navy is a helicopter operator and has two squadrons flying ASW, minesweeping and rescue duties. For these missions there are 10 Boeing/Kawasaki KV-107s, 10 Alouette IIs and 10 Agusta-Bell AB206B JetRangers. The Navy has taken over Air Force rescue duties and has incorporated into its fleet the SAF machines comprising 10 Boeing-Vertol 107s and six AB204Bs.

The Army Aviation Department provides AOP and communications duties equipped with five Dornier Do27s, 12 Piper Super Cubs and 20 SA Bulldogs. Helicopters used number 12 AB204Bs, 40 AB206As and six Alouette IIs.

Switzerland

Title: Swiss Air Force and Anti-Aircraft Command
Headquarters: Bern
Strength: 650 aircraft

Combat

A part of the Army, the Swiss Air Force has about 300 combat aircraft and some 45,000 personnel, of

Above: *Swiss Air Force DH Venom F Mk 4 fighter-bomber.*

which a large percentage are part-time militiamen. An efficent system of rock shelters and associated runways is operated to protect its aircraft and equipment in time of emergency. Sections of Swiss roads are also designated runways in time of war. The force currently incorporates all the operational units into three regiments (Fleigerregimenter) each consisting of between six and eight squadrons, or Fliegerstaffeln, with an establishment of some 18 aircraft each.

In 1979, the first batch of Northrop F-5E air-superiority fighters will join the front-line Surveillance Wing of the Air Force, which currently operates three HS Hunter squadrons, two Dassault Mirage squadrons and a reconnaissance unit with Mirages and DH Venoms. Ordered in 1976, a total of 72 F-5s are destined for service with four squadrons (Nos 1, 11, 18 and 21) comprising 66 single-seat Es and six two-seat Fs, the majority being assembled at the Federal Aircraft Factory, Emmen, from components air-freighted from the USA. Just before selection of the F-5E, and as an interim measure, the air force purchased a further 60 Hunters in 1974. These aircraft, Mk 58As and eight two-seat Mk 68s, together with existing machines in the inventory took the SAF Hunter force to 148 aircraft. Armed with Sidewinder missiles and Saab BT9K bombing computers, the Hunters equip nine squadrons (Nos 1, 4, 5, 7, 8, 11, 18, 19 and 21) in the ground-attack role, with three squadrons (Nos 1, 11 and 18) assigned to the Surveillance Wing.

Approximately 100 DH Venom F Mk 4 fighter-bombers of 250 originally in service equip seven squadrons (Nos 2, 3, 6, 9, 13, 15 and 20). The Dassault Mirage force is made up of 36 IIIS interceptors and two IIIBS trainers flying with Nos 16 and 17 Squadrons at Emmen and Payerne, integrated into the Hughes Florida early-warning and air-defence system. Entering service in 1966, the 57 Mirages procured included 16 IIIRS tactical reconnaissance variants, and these fly in No 10 Squadron alongside eight camera-equipped Venoms.

Transport and Support

The Swiss Air Force maintains a transport flight of three immaculate Junkers Ju52/3m tri-motor transport aircraft, these conducting paratroop and freight duties. Seven light-aircraft squadrons (Nos 1-7) fly liaison, communications and SAR missions using three Beech Twin Bonanzas, six Dornier Do27s, 11 Pilatus Porters and six Turbo-Porters, with a helicopter element equipped with 27 Alouette IIs and 80 Alouette IIIs. A target-towing unit has 23 Federal C-3605 turboprop-powered tugs introduced into service in the early 1970s.

Basic training is conducted on AS202 Bravos, pupils continuing on to the Pilatus P-2/P-3, of which there are some 120 in service. Jet training is flown on about 40 DH Vampire Mk 6s and 35 Vampire T55s flying in the advanced role.

Republic of Taiwan

Title: Chinese Nationalist Air Force
Headquarters: Taipei
Strength: 734+ aircraft

Combat

A recipient of large-scale American aid, Taiwan has a powerful air force and has established an aircraft

manufacturing industry. Known as the Aero Industry Development Centre and situated at Taichung in Central Taiwan, it is assembling an initial batch of 120 Northrop F-5Es for the CNAF against an air force requirement for some 300 to replace older types currently in service. A total of 173 F-5Es and 27 F-5F two-seat combat trainers are on order to date, 60 being supplied direct from the USA. Another new type for which the Taiwan Government was negotiating for in mid-1978, was the Israeli IAI Kfir C-2 fighter-bomber; the US approved the sale of 50 aircraft equipped with the American made J79 engine, but subsequently Taiwan turned down the

Top: *About 40 Fairchild C-119G Packets are in service with the Taiwan Air Force.* / H. Holmes

Above: *Chinese Nationalist Army Bell UH-1H.* / Bell

offer. However, Taiwan has purchased quantities of the Israeli Shafrir air-to-air missile for CNAF use.

The air force is organised into wings, each made up of three squadrons with between 18 and 25 aircraft. Three fighter squadrons make up the 1st Fighter Wing at Kung Kuan equipped with 70 Northrop F-5A/Bs of 92 F-5As and 23 F-5Bs supplied in the 1960s, while the 2nd and 3rd Wings are relinquishing nearly 100 obsolete F-86F Sabres for the new F-5Es. The 4th Fighter Wing has three squadrons with about 80 North American F-100A/D Super Sabres of a total of 114 received, and the 5th FW has Lockheed F-104G Starfighters numbering 63 aircraft plus a reconnaissance unit with eight RF-104Gs. Attached to the Taiwan Navy is an MR and ASW unit flying nine Grumman S-2A Trackers supported by a rescue unit with a few Grumman HU-16B Albatross amphibians.

Transport and Support

More than 120 twin-engined aircraft make up the transport fleet comprising a Wing of 40 Fairchild C-119G Packets and 10 C-123 Providers; also in use are 30 Curtiss C-46 Commandos and 50 Douglas C-47s plus about six Douglas C-54s. There is a Boeing 720B for VIP use while the helicopter fleet includes six Hughes 500s, seven Sikorsky UH-19s, 10 Bell 47Gs and more than 60 AIDC-assembled Bell UH-1Hs.

Another AIDC product is the PL-1B Chienshou primary trainer of which 50 have been built, together with a turboprop powered version of the T-28 known as the T-CH-1B, 30 being planned. Both these aircraft are replacing T-6s and T-28s. For jet training about 30 Northrop T-38s and some T-33As are flown, while two-seat F-5Bs, TF-104Gs, F-100Fs and F-5Fs are attached to the respective units.

The Chinese Nationalist Army has an aviation element for support, troop transport and liaison work. Helicopters form the equipment of this force made up of 50 locally-assembled Bell UH-1Hs delivered in 1973-75, seven Sikorsky CH-34s, and two Kawasaki KH-4s.

Syria

Title: Syrian Arab Air Force
Headquarters: Damascus
Strength: 500 aircraft

Combat

Backbone of this Soviet-equipped air arm is a force of some 250 MiG-21PF/MF/bis interceptors operating in 12 squadrons divided into four regiments. There is also a single regiment of three squadrons equipped with about 45 variable-geometry MiG-23/27 Floggers, Syria being one of four Arab states to be supplied with these interceptor/strike aircraft. Doubling in the low-altitude fighter-bomber role are about 50 MiG-17Fs, with a further 60 Su-7s and some Su-20 swing-wing strike aircraft equipping the ground-attack elements.

Transport and Support

The transport force has eight Il-14s, six Douglas C-47s, six An-12 freighters, four Il-18s, and two Piper Navajos for survey and liaison work. At least nine Kamov Ka-25s are used for coastal ASW patrols, while as many as 50 Mi-8s are operated on assault and transport duties together with a lesser number of Mi-4s and Mi-6 heavy-lift helicopters. Helicopters figure prominently in Syria's purchases outside the Soviet Bloc. From Italy, the Air Force is receiving six Meridionali-built Boeing CH-47C Chinooks, 18 Agusta-Bell AB212s and 12 Agusta-built Sikorsky SH-3D Sea Kings, while from France at least 16 SA342L Gazelle liaison machines are on order, some equipped for the anti-tank role.

Trainers include Yak-11s, Yak-18s, L-29 Delfins, and 48 Siat Flamingo basic trainers bought from Spain and Germany. Operational conversion is performed on two-seat versions of the front-line combat aircraft.

Tanzania

Title: Tanzanian People's Defence Force Air Wing
Headquarters: Dar-es-Salaam
Strength: 72 aircraft

The combat element of the Tanzanian Air Wing is composed of three squadrons of Chinese-supplied jets totalling 15 MiG-21Fs (Shenyang F-8), eight MiG-19s (Shenyang F-6) and 10 MiG-17Fs (Shenyang F-4). Two MiG-15UTI trainers have also been received, and the main operating base for the MiG element is Mikumi, situated near Dar-es-Salaam.

The Transport Group operates a mainly western-equipped force, the most recent arrivals being three HS748 passenger/freight aircraft ordered in 1977, and six DHC-5D Buffalo STOL tactical transports. These join 12 DHC Caribous, six Cessna 310s (including two 310Qs) and a single An-2 biplane. A basic training element has five Cherokee 140s and there is a civilian-registered HS748 operated for VIP flights. The small Police Air Wing helicopter element has two Bell 47Gs and two AB206s, fixed-wing types amounting to a Cessna U206 and a Piper Cherokee Six.

Thailand

Title: Royal Thai Air Force
Headquarters: Bangkok
Strength: 450+ aircraft

Combat

To increase the effectiveness of the RTAF, which is currently involved in an anti-guerrilla war, new combat aircraft are being procured, mainly from the United States. Due for delivery late in 1978, is the first of 17 Northrop F-5E fighter-bombers and three F-5F two-seat trainers ordered in 1976 at a cost of $50 million. These will join the force of older F-5As delivered in 1966, comprising No 13 Squadron at Don Muang, Bangkok, equipped with 24 aircraft plus two F-5B trainers and four RF-5A reconnaissance aircraft. With No 13 Squadron in the 1st Combat Wing are No 11 Squadron with 20 Lockheed T-33As and four RT-33As, and No 12 Squadron, with some of the 32 Rockwell OV-10C Bronco counter-insurgency aircraft supplied to Thailand in two batches of 16 aircraft from 1971 onwards; a further six Broncos were announced as ordered for the RTAF in 1977 by the US Government, presumably to make up for attrition.

The 2nd Wing has at least six counter-insurgency squadrons (Nos 21, 22, 23, 53, 62 and 73) equipped with the remainder of the OV-10Cs, 30 North American T-6Gs, 16 Cessna A-37Bs, 45 North American T-28Ds, and 31 Fairchild Hiller AU-23A Peacemakers. Most of the above units operate within the 2nd, 21st, 23rd and 41st Air Wings under a recent reorganisation of the RTAF's combat elements.

Transport and Support

Based at Bangkok and forming No 6 Wing, are Nos 61 and 62 Squadrons, operating between them 20 Douglas C-47s, more than 40 Fairchild C-123B/K Providers and five Beech C-45s. Two HS748s are attached to the Royal Flight which also has two Swearingen Merlin IVAs delivered in December 1977 and March 1978. A single helicopter squadron, No 63, operates most of the RTAF rotary-wing force of 63 Bell UH-1Hs, 40 Sikorsky H-34Cs, 13 UH-19s and three Kaman HH-43B Huskies. Phased into service in 1978 was a batch of 18 ex-US Army H-34s modernised to S-58T standard by the Thai-Am concern in Bangkok.

Training is centred at Korat Air Base, with basic work performed on 24 CT/4 Airtrainers, 12 Siai-Marchetti SF260MTs, at least four Cessna T-41Ds and 10 Continental engine-powered DH Chipmunks. Advanced training is flown on 14 Cessna T-37Bs but the Air Force is known to be keen to procure a quantity of Aermacchi MB326s, not only for use in the training role but also as armed light strike aircraft. For liaison purposes there are Helio U-10A Couriers, Beech U-8F Queen Airs and some DHC Beavers.

The Royal Thai Navy has a modest airborne element, based in Bangkok, and operated in the ASW and SAR roles. Ordered in 1977 and delivered a year later, were two Canadair CL-215 amphibians procured for the rescue task and believed destined to replace two Grumman HU-16B Albatross. In addition

Below: *Standard advanced trainer with many air arms is the Lockheed T-33A; this one is a Thai example.*

Above: *One of two Thai Navy Canadair CL-215 amphibians.*

there is a maritime reconnaissance unit equipped with 10 Grumman S-2A Trackers delivered in 1966.

The Royal Thai Army has an airborne element with more than 200 machines, both fixed-wing and helicopters, flown mostly in support of ground forces. For observation and liaison there are 90 Cessna 0-1s and a single Beech 99 for transport duties; helicopters include about 90 Bell UH-1B/Ds, 16 FH-1100s, six Hiller OH-23Fs, three Bell 206s, and four Boeing CH-47A heavy-lift Chinooks. The para-military Thai Border Police have three DHC Caribous, three Short Skyvans, five Fairchild Peacemakers, four Pilatus Porters, three Dornier Skyservants and one CT/4 Airtrainer. A helicopter element has 10 Bell 204Bs, 11 Bell 205s, two Bell 205As and four 206Bs.

Togo

Title: Force Aérienne Togolaise
Headquarters: Lomé
Strength: 22 aircraft

This small African air arm, formed following independence from France in 1960, has expanded over the past few years from a purely transport and supply force to one having a modest strike potential. This attack capability is based on six Embraer EMB326GC Xavante two-seat strike/trainers procured from Brazil in two batches in 1976 and 1978. In addition, five Alpha Jets have been ordered

Below: *Togo Air Force DHC-5D Buffalo.* / DHC

from France, following the purchase of five Fouga Magisters in 1975 for advanced training.

The transport arm has increased its inventory with two DHC-5D Buffalo freighters which arrived in 1976, joining the existing force of two Douglas C-47s, two MH1521M Broussards and two Reims Super Skymasters. There is also an SA330 Puma, a Lama and a single Fokker-VFW F28 Mk 1000, the latter for VIP use.

Tunisia

Title: Tunisian Republican Air Force
Headquarters: Tunis
Strength: 70 aircraft

Financial considerations in 1976 prevented the Tunisian Government from accepting a letter of offer from the United States for 16 Northrop F-5E fighter-bombers. These were to have replaced the survivors of 12 North American F-86F Sabres supplied by the USA in 1969, but subsequent events have kept the Sabres in service for training purposes while the TRAF turned to less costly combat equipment. Having purchased eight Aermacchi MB326Bs in 1965 for strike/training, the Service ordered a further batch in 1976 comprising eight single-seat MB326KTs and four two-seat MB326LTs, and these equip a dual counter-insurgency/trainer unit. For basic tuition there are 12 Siai-Marchetti SF260WTs and six SF.260Cs ordered in 1978 plus 12 North American T-6 advanced trainers. Three Dassault Flamants, eight Alouette IIs, six Alouette IIIs and at least one SA330 Puma provide transport and liaison.

Below: *Tunisian Air Force SF260WT Warrior used for armed training.*

Turkey

Title: Turk Hava Kuvvetleri
Headquarters: Ankara
Strength: 700+ aircraft

Combat

Turkey is a key member of NATO and her air force operates as part of the 6th Allied Tactical Air Force. The country's dispute with Greece over Cyprus in 1974 forced America to cut arms supplies to Turkey which were only reinstated in 1978. Prior to this limited amounts of equipment were received, but as a result of the mid-seventies embargo, new equipment orders are likely over the next couple of years. The Turk Hava Kuvvetleri is divided into three main components under the overall direction of the Air Force Central Command. The combat squadrons operate under the command of the First Tactical Air Force, Eskisehir Air Division, and the Third Tactical Air Force, Diyarbakir Air Division. The third component is Training Command, Izmir Air Division. From 12 air bases plus a further 17 smaller airfields used in emergencies, the air force operates a total of

Above: *Aeritalia-built F-104S Starfighter of the Turkish Air Force.* / Aeritalia

22 combat squadrons with more than 450 aircraft. McDonnell Douglas F-4E Phantoms currently equip two strike squadrons with a total of 40 aircraft; a further 32 F-4Es and four RF-4E reconnaissance Phantoms are being delivered for two more units. Forty Aeritalia-built F-104S Starfighters equip two multi-role squadrons, supplementing two more strike units flying 32 Lockheed F-104Gs and four TF-104Gs, and a third squadron operating 20 ex-Italian AF F-104Gs bought in 1976.

Both Greece and Turkey were supplied with Convair F-102As by the United States and the latter continues to operate the type although in diminishing numbers as new Phantoms arrive. Two interceptor squadrons have 35 ex-USAF single-seat F-102As plus three TF-102A trainers. Ground-attack missions are performed by six squadrons operating 125 Northrop F-5As, 15 F-5B trainers, and seven ex-Libyan F-5As to make up for attrition. A reconnaissance force of two squadrons has 36 RF-5As, and a further strike element of five squadrons has a total of 80 North American F-100D/F Super Sabres.

Transport and Support

The main transport base is Kayseri and there are four squadrons providing tactical and long-range transport duties. Heavy-lift work is flown by No 222 Squadron at Erkilat with seven Lockheed C-130E Hercules and No 221 Squadron at the same base with 20 C160 Transalls, while other tasks are flown by 14 Douglas C-47s, three C-54s, and six Dornier Do28s; three Britten-Norman Islanders are in use for survey flights. Helicopters are used for liaison and support, there being some 40 Agusta-Bell AB204/205s in service.

Training Command, with a main base at Cigli in Izmir, has a selection of types taking pupils from the primary and basic stages to advanced flying, the inventory including 19 Cessna T-41s flying with No 123 Squadron at Cumaovasi. five Beech T-42 twin trainers, 23 Cessna T-37B/Cs and 30 Lockheed T-33As. A small number of Cessna 421Cs have recently been delivered to the service.

The Turkish Navy has an airborne arm tasked mainly with anti-submarine and reconnaissance. A Grumman Tracker fleet comprises eight ex-Dutch Navy S-2As delivered in 1971, 12 ex-US Navy S-2Es and two TS-2A trainers, while a helicopter element has 12 Agusta-Bell AB212ASWs with a further six on option, plus three AB205s for liaison duties.

The Turkish Army flies an assortment of types in support of ground forces, divided into flights at divisional and corps level. Helicopters include 140 AB204/205, 15 AB206s and 20 Bell 47Gs; a further 56 AB205s are being delivered in two batches of 28 aircraft each. Fixed-wing types are made up of nine Dornier Do28 transports, 15 Do27 liaison machines, 18 Cessna 421Bs, five Beech Barons and seven Piper L-18 AOP aircraft.

Uganda

Title: Uganda Army Air Force
Headquarters: Entebbe
Strength: 71 aircraft

Since the destruction of half of UAAF's combat force in the July 1976 Israeli commando raid on Entebbe, the Soviet Union has replaced the 11 MiGs lost, and the present Ugandan combat inventory now comprises two squadrons, one with 12 MiG-17Fs and the other with 10 MiG-21s. Supporting these units is a third assigned the light strike role and flying eight Magisters originally supplied by Israel in the 1960s. For training there are at least two MiG-15UTIs and five Czech L-29 Delfins in the advanced

role, while for basic tuition there are five Piaggio P149s. Six Swiss AS202 Bravos have been received for primary training with the Central Pilot School.

A single Grumman Gulfstream II is in service for VIP duties and there is a DHC Twin Otter and a Caribou for transport work, both flown by the Ugandan Police Air Wing. A helicopter element has six Agusta-Bell AB205s and four AB206s, plus a number attached to the PAW. The Uganda Army has about 10 Piper Super Cubs for liaison, training and spotting.

Above: *Uganda Army Air Force Piper Super Cub used for liaison duties.* / S. R. P. Thomson

United Arab Emirates

Title: United Emirates Air Force
Headquarters: Abu Dhabi
Strength: 96 aircraft

Combat

The UEAF was formed by the amalgamation of the three air units of Abu Dhabi (AD Air Force) and Dubai (Dubai Police Air Wing and the Union Air Force). Seven Gulf States make up the UAE federation — Abu Dhabi, Amman, Dubai, Fujairah, Ras al-Khaimah, Sharjah and Umm al-Qaiwain — and all contribute to the funding of the Air Force, which is headed by UAE President Sheikh Zayed of Abu Dhabi. The UAE has its headquarters at Abu Dhabi but a Central Air Force base has been established at Dubai Airport. Providing an air defence nucleus is a wing of 32

Below: *One of four DHC-5D Buffaloes for the United Arab Emirates delivered in 1978.*

Dassault Mirages based at Abu Dhabi. The first of these — 12 5AD strike fighters and two 5DAD combat trainers — arrived in 1974, but further deliveries in 1976-77 have taken the force to 26 Mirage 5AD/EADs, three reconnaissance 5RADs and three 5DAD two-seaters flying in two squadrons. Assisting UAE nationals in operating these aircraft are Pakistan Air Force personnel.

For ground-attack duties there is a squadron of HS Hunters based at Sharjah comprising eight FGA76s and two T77 trainers, while at Dubai's Central Air Force base is a small counter-insurgency force, part of the Police Air Wing, equipped with three armed single-seat Aermacchi MB326KDs and a two-seat MB326LD. Four more of these strike/trainers have been ordered and were expected to be delivered by the end of 1978.

Transport and Support

The transport element is based at Abu Dhabi but operates throughout the UAE. Largest type in service is the Lockheed C-130H Hercules, two of which were delivered in 1975, and these are supplemented by five DHC-5D Buffalo STOL transports received in 1978, three DHC Caribou and four Britten-Norman Islanders. Helicopters in use total ten Aérospatiale SA330 Pumas, seven Alouette IIIs and some Italian AB205As, but this fleet is likely to be expanded in the early 1980s when Egyptian Westland Lynx production gets underway. The Arab Organisation for Industrialisation (AOI) and Westland/Rolls-Royce signed the Lynx manufacturing agreement in 1978 and apart from Egypt, the other recipients of the 230 machines will be Saudi Arabia and the UAE.

Above: *Dubai, forming part of the UAE, operates AB206Bs.* / Bell

At Dubai there is a single Aeritalia G222 STOL transport delivered in December 1976 and a second machine is on option. The three ex-Abu Dhabi AB206Bs which formed the basis of the Union Air Force in 1972 have since been joined at Dubai by four Bell 205As used for troop transport work, and three AB212s. The Police Air Wing has three AB206Bs and two Bell 205As, used for liaison and communications. There is also a Cessna 182 for training and an SF260WD Warrior for armed training. Miscellaneous types include a Piper Pawnee Brave crop-spraying aircraft delivered in December 1975, and a Lake Buccaneer amphibian for the personal use of Sheikh Kalifa, both machines being based at Abu Dhabi.

United States of America

Title: United States Air Force
Headquarters: Washington
Strength: 9,200 aircraft

Combat

The USAF organisation is based on Wings comprising up to three squadrons each with an establishment of between 10 and 24 aircraft. One of the two largest air forces in the world, the USAF has eight major operational commands: Strategic Air Command, Aerospace Defence Command, Tactical Air Command, US Air Forces Europe, Pacific Air Forces, Alaskan Air Command, Military Airlift Command and Air Training Command.

Supplementing the major commands are Air National Guard and Air Force Reserve units.

Strategic Air Command maintains the USAF deterrent and all nine ICBM wings and the manned bomber force comes under its control within the 8th and 15th Air Forces. The bomber element has 19 wings of Boeing B-52s, comprising 269 B-52G/Hs (many equipped with SRAM short-range air-to-ground missiles and fitted with an electro-optical viewing system) and 80 structurally modified B-52Ds. Only one B-52 unit is now permanently based overseas, the 43rd Wing on Guam; approximately 20 home-based B-52Fs equip one training unit. SAC began sea-surveillance flights in 1976 in co-operation with the US Navy, in addition to their regular long-range strategic missions. Each of the bomber wings has attached to it an associated Boeing KC-135 Stratotanker element. Formed into 35 squadrons with 615 aircraft of the total original KC-135 procurement of 732, these also support other commands on air-refuelling duties and include 80 aircraft assigned to the Air Force Reserve. Four squadrons in two wings, the 380th and 509th, operate 70 SRAM-equipped FB-111As, while SAC's reconnaissance force comprises the 9th Strategic Reconnaissance Wing with nine Lockheed SR-71s and its supporting KC-135Q tankers at Beale AFB. Two further wings with 20 Lockheed U-2s, nine DC-130s (for drone launching), and about 50 Boeing RC-135s and EC-135s fly a variety of tasks, including electronic reconnaissance, atomic sampling, etc. In addition, 14 Air Guard units fly KC-135s assigned to SAC duties. Following the cancellation of the Rockwell B-1 variable-geometry strategic bomber in June 1977 (241 B-1s were planned to replace the B-52 force), the effectiveness of the present force of B-52s is to be increased with the procurement of up to 1,000 cruise missiles. Other new types joining the force or scheduled to arrive during the next few years include 25 Lockheed TR-1 surveillance machines to supplement the U-2s, 20 McDonnell Douglas KC-10A Advanced Tanker/Cargo Aircraft, up to 34 Boeing E-3A Airborne Warning and Control System aircraft of which three R&D and 16 production aircraft have so far been funded, and six Boeing E-4 Command Post versions of the Boeing 747 airliner.

Above: *The West's largest lifter — the USAF Lockheed C-5A Galaxy.*

Aerospace Defence Command, while regarded more as a missile warning force, still maintains its important air-defence commitment over the continental US with the co-operation of the Canadians under the North American Air Defence Command (NORAD). The interceptor force comprises six active squadrons with General Dynamics F-106s, five Air Guard squadrons with F-106s, three squadrons with F-4 Phantoms (including one based in Iceland and one in Hawaii), and three ANG squadrons with 60 McDonnell F-101B Voodoos. There are two Groups of Martin EB-57s with the ANG and a further squadron in the active inventory. The ADC's remaining 10 Lockheed EC-121T Warning Stars have been assigned to the Air Force Reserve and fly from Florida and Iceland.

Tactical Air Command is the USAF's quick-reaction force, incorporating fighter, reconnaissance, transport and special operations units. Approximately 1,700 aircraft operate within TAC and its two Air Forces, the 9th with headquarters at Shaw AFB, SC, and the 12th headquartered at Bergstrom AFB, Tex. Following a year with the 355th TF Training Wing at Davis-Monthan AFB, Fairchild A-10A Thunderbolt IIs are now fully operational with the 354th TFW at Myrtle Beach, funding having been allocated for 339 aircraft up to 1978. Plans call for a total of five wings of A-10s, each with four 18-aircraft squadrons with the total USAF requirement standing at 733 machines. McDonnell Douglas F-15 Eagles are operational with the 1st TFW at Langley AFB and the 49th TFW at Holloman with the 33rd TFW at Eglin converting from Phantoms to Eagles for full operational status by October 1979. A total of 729 Eagles and 20 development aircraft are scheduled for USAF service with a total of 19 squadrons. The first General Dynamics F-16 Wing will be the 388th TFW at Hill AFB which will receive its first aircraft early in 1979, a total of 1,388 of these Air Combat Fighters being required by the Air Force of which 206 will be two-seat F-16Bs. Present types in use include two active AF squadrons with 44 Republic F-105G 'Wild Weasel' Thunderchiefs, eight squadrons with 210 LTV A-7D Corsair IIs, seven GD F-111E/F squadrons with 180 aircraft, seven McDonnell Douglas RF-4C squadrons with 121 aircraft, and 30 squadrons with 664 F-4C/D/E Phantoms. Seven special-operations squadrons fly 42 Rockwell OV-10As, 68 Cessna 0-2A/Bs, 27 Lockheed C/AC-130s and five EC-135s. Most of TAC's 82 Northrop T-38s and 55 F-5Es are used for 'Red Flag' combat training at Nellis AFB. Total TAC helicopter inventory is 19 Bell UH-1s, 15 Sikorsky CH-3s and four CH-53s. Air National Guard units attached to TAC include three wings and five groups of A-7D Corsair IIs, four wings and eight groups of North American F-100D Super Sabres, two wings and two groups of F-105B/Ds, two Cessna A-37B groups, two wings and six groups of F/RF-4Cs, one RF-101 group, and two wings and four groups of Cessna 0-2As.

US Air Forces Europe is a part of NATO, with some 28 squadrons based with the 3rd AF in the UK, 16th in Spain and 17th in Germany. Future plans for USAFE call for three wings of Fairchild A-10As to be based in Europe, one (81st TFW at Bentwaters with six squadrons) to be based in England and two in Germany; first deliveries were in January 1979. The second unit to get F-15 Eagles is the 32nd TFS at Soesterburg which received 18 aircraft in 1978, following the 36th TFW at Bitburg which has 72 aircraft. Most numerous type in service is the Phantom, of which some 320 equip four wings of 11 squadrons belonging to the 401st TFW, Torrejon, Spain; 50th TFW, Hahn, Germany; 52nd TFW, Spangdahlem, Germany; 86th TFW, Ramstein, Germany). Other major USAFE units include one wing of three squadrons of 80 GD F-111Es (20th TFW, Upper Heyford, UK), one wing of three squadrons of 80 F-111Fs (48th TFW, Lakenheath, UK), two reconnaissance wings with 60 RF-4Cs (one squadron with the 10th TRW at Alconbury, UK, and two squadrons with the 26th TRW at Zweibrucken, Germany), and a rotational squadron of some 15 Lockheed C-130E/H Hercules detached to Mildenhall, UK. At the same base are a number of SAC rotational KC-135s. For Aeromedical duties four McDonnell Douglas C-9A Nightingale transports are based at Rhein-Main, Germany, together with another rotational Hercules unit from the USA. At Sembach 40 Rockwell OV-10A are deployed with the 601st Tac Control Wing, which also has seven CH-53Cs. Located at Woodbridge, UK, the 67th ARRS is controlled by MAC but provides helicopters and aircraft for rescue duties throughout the northern hemisphere, four Sikorsky HH-53Cs and seven HC-130H/Ns forming the unit's equipment. Supporting USAFE are a number of smaller liaison units flying Rockwell T-39s, Beech C-12As, Lockheed VC-140Bs, Bell UH-1Hs and UH-1Ns.

Pacific Air Forces, with its headquarters in Hawaii, comprises the 5th AF based in Japan and the 13th AF in the Philippines. Main units are as follows: under 5th AF command, 8th TFW at Kunsan AB, Korea, with F-4s; 18th TFW at Kadena AB, Okinawa, with 90 F-4/RF-4Cs; and the 51st Composite Wing at Osan AB, Korea, with F-4s and OV-10As. The 3rd TFW is based at Clark AB, Philippines, and flies F-4s and a squadron of F-5Es designated 26th TF Training Squadron, while at Hickham itself a support wing has a number of EC-135s, Lockheed T-33As and Cessna 0-2As.

Alaskan Air Command operates from Elmendorf AB and has a Lockheed C-130E transport squadron (17th TAS), an F-4E Phantom squadron with 26 aircraft assigned to ADC (43rd TFS), an HH-3/HC-130 squadron for SAR duties (71st ARRS) and a tactical operations unit with T-33As, EB-57s and a T-39 (5041st TOS). Also operated is the 25th Tac Support Squadron with 0-2As at Eielson AB and an Air Guard unit with eight C-130 Hercules (176th TAG).

Transport and Support

In addition to the units assigned and operated within the separate commands listed above, the major USAF transport force comes under the wing of Military Airlift Command which also encompasses the Air Weather Service, the Aerospace Rescue and Recovery Service and the Aeromedical Airlift Wing. Two Air Forces, the 21st and 22nd incorporate 13 squadrons with 271 Lockheed C-141 Starlifters and four squadrons with 76 Lockheed C-5A Galaxy heavy-lift freighters. Six other squadrons have 23 C-9A Nightingales (including three VC-9Cs with the 89th MAW), 11 VC-140s, two VC-137s, three VC-137Bs, four C-131s, 11 C-135s and 14 WC-130s. The large tactical transport force of C-130E/H Hercules has some 276 aircraft flying with 14 active AF units, while the supporting Air Guard inventory has 13 groups and five wings of C-130s. Also MAC-assigned is an ANG group of 19 DHC C-7 Caribous and two groups flying HH-3/HC-130 rescue aircraft. Three T-39 training/liaison units are attached to the 89th MAW with 105 aircraft, and being delivered are 34 Beech C-12s for the use of air attaches around the world. Helicopters include more than 120 Bell UH-1Fs for missile site support, 30 UH-1Hs for base rescue duties, 51 HH-1Ns, 46 Sikorsky HH-3Es, and 33 of 72 HH-53B/Cs in USAF service.

Air Training Command with headquarters at Randolph AB, Tex, operates a total of 1,629 aircraft, comprising 692 Cessna T-37s, 822 Northrop T-38s, 96 Cessna T-41s and 19 Boeing T-43s. The Navigator Training Wing at Mather AB, Calif, operates as a joint USAF/USN/USMC/Coast Guard unit with the T-43s. At the USAF Academy, Colorado Springs, two DHC UV-18B Twin Otters provide platforms for cadet parachute training. The Air Force

Above: *US Marine Corps' Grumman A-6 Intruders.*

Reserve is a large unmobilised training and support element with headquarters at Dobbins AB, Ga, and it flies many of the roles of the active AF. Reserve units are divided into three Air Forces (4th, 10th and 14th AF) within the US, flying 11 C-130 squadrons, two C-7 squadrons, four Fairchild C-123K squadrons, two KC-135 squadrons, three F-105 squadrons, five A-37B squadrons, one AC-130 squadron, four ARRs squadrons with HH-1H, HH-3E and HC-130 aircraft, one CH-3E unit and a weather squadron with WC-130s.

Title United States Navy
Headquarters: Washington
Strength: 2,500+ aircraft

Numerically the world's third largest air force, the US Navy deploys the great majority of its combat force on 13 giant attack carriers, seven of them operating with the Atlantic Fleet, based at Norfolk, Va, and six with the Pacific Fleet based at Alameda and San Diego, Calif. The latest ship to join the carrier force was the nuclear-powered *Dwight D. Eisenhower* which joined the *Enterprise* and *Nimitz* in 1978 as the third of four nuclear carriers planned; the fourth ship is the *Carl Vinson* which is due for service in 1981. When deployed, each of the attack carriers embarks a Carrier Air Wing totalling between 80 and 95 aircraft divided into the following units: two interceptor squadrons with 11 McDonnell Douglas F-4B/J Phantoms or Grumman F-14A Tomcats, two attack squadrons each with 10 LTV A-7E Corsairs, a medium attack squadron with Grumman A-6E and KA-6D Intruders, a detachment from one of five reconnaissance squadrons flying North American RA-5C Vigilantes, a detachment of four Grumman EA-6B Prowlers from an electronic warfare squadron, an AEW squadron with four Grumman E-2B/C Hawkeyes, one ASW squadron with 10 Lockheed S-3A Vikings, and one helicopter ASW squadron with eight Sikorsky SH-3D Sea Kings. Carrier philosophy in recent years has seen the attack class take on the specialised ASW duties which previously had been handled by special ASW carriers, hence the inclusion of the Sea Kings and Vikings in the average carrier complement.

The US Navy has nearly 1,900 combat aircraft alone, flying in 28 interceptor squadrons (F-4/F-14), 41 attack squadrons (27 with A-7s, 12 with A-6s and two with A-4s) and six reconnaissance squadrons with RA-5s and one with RF-8G Crusaders. Future procurement for this front-line inventory includes a planned purchase of 800 McDonnell Douglas F-18 Hornet Naval Strike Fighters, the prototype of which made its first flight late in 1978; the US Marine Corps is to receive some of these aircraft to replace its Phantoms from 1982. Plans call for six active and four reserve USN squadrons to receive 185 F-18s, and 345 to replace A-7s from the mid-1980s. Grumman F-14A Tomcat deliveries continue against a total procurement of 521 aircraft for 18 squadrons; some F-14As are being converted to the reconnaissance role. A total of 65 E-2A/B Hawkeyes have been delivered to the service, and there are 77 E-2Cs delivered or on order, while all 187 S-3A Vikings ordered have been received by the 12 ASW units assigned the type. A

similar number of units are receiving EA-6B Prowlers, a total of 77 currently (1978) being on order.

Land-based long-range maritime patrol duties are performed by 24 squadrons in five patrol wings with about 280 Lockheed P-3A/B/C Orions of some 428 delivered to the Service. Twelve EP-3Es are flying in the recce role, together with some Convair EC-131s and Lockheed EC-130s. There are seven helicopter support squadrons with Kaman UH-2s, Boeing-Vertol UH-46s and HH-2Ds, and mine-countermeasures unit with 30 Sikorsky RH-530s. A higher performance, heavy-lift version of this large helicopter, designated CH-53E, is being built for the Navy, 18 being on order. Another new type and winner of the USN Light Airborne Multi-Purpose System competition, is the Sikorsky SH-60, for which the service has a requirement for 204 machines, the prototype was scheduled to make its first flight early in 1979.

Ten Sikorsky SH-3D Sea King squadrons have some 80 aircraft, while six transport squadrons provide worldwide fleet support equipped with 30 Douglas C-118s, seven C-130Fs, 12 C-9B Nightingales, 12 CT-39s, 20 C-1 Traders and 12 C-2 Greyhounds. Late in 1977, the USN ordered 22 Beech UC-12Bs for liaison duties and as many as 66 could be purchased.

Twenty-one training squadrons in six wings have a wide variety of types for the various training tasks undertaken. For basic tuition 278 Beech T-34C Turbo-Mentors are replacing older T-28s and T-34s, while advanced flying is performed on North American T-2C Buckeyes in five squadrons with pupils finishing the course on two-seat TA-4J Skyhawks currently equipping six squadrons. Multi-engine training is flown on 65 Beech T-44A King Airs, replacing T-2B/Cs. Sixty TA-7Cs are on order for advanced training, the airframes being converted from existing A-7B/C single-seat Corsairs. A Fighter Training Aggressor Squadron is equippped with 10 Northrop F-5Es and three F-5F two-seaters performing combat training simulating Eastern Bloc tactics. Helicopter training is flown on Bell TH-57A SeaRangers to begin with, followed by advanced flying on Bell TH-1Ls and UH-1E/Hs.

Above: *Carrier-based F-14A Tomcat of VF-142, US Navy.* / B. C. Wheeler

In addition to the active units, the Navy Reserve Force has four F-4N interceptor squadrons, six A-7 attack squadrons, one RF-8G reconnaissance squadron, one E-2 Hawkeye AEW squadron and one KA-3B Skywarrior tanker squadron. Seven helicopter squadrons in the Reserve have SH-3s and HH-1s, while 13 reserve land-based patrol squadrons operate early versions of the P-3 Orion.

Title: United States Marine Corps
Headquarters: Washington
Strength: 800 aircraft

This compact air force supports Marine ground troops wherever they operate and is divided into three Marine Air Wings (1st MAW at Okinawa, 2nd MAW at Cherry Point, 3rd MAW at El Toro) with the 4th MAW at Glenview controlling reserve forces. Most numerous type in the USMC inventory is the F-4 Phantom, and 12 fighter-bomber squadrons are equipped with 144 F-4B/Js which have been converted to F-4N/S standard. These aircraft are due to be replaced from 1982 by 270 F-18 Hornets. In the light attack role there are five squadrons with about 70 A-4F/M Skyhawks, and three squadrons and a training unit flying 110 AV-8A Harriers, including eight two-seat TAV-8As. A total of 336 AV-8Bs is required by the USMC to replace the -8As and A-4s and development work on the project continues. Five all-weather attack squadrons have 60 A-6A/E Intruders, while three reconnaissance squadrons operate a total of 42 EA-6As and RF-4B Phantoms. The Corps has 15 EA-6B Prowlers on order to replace the EA-6As which are being passed to the USN.

The Marine transport force has three squadrons equipped with 46 Lockheed KC-130F Hercules tanker/transports; up to 14 updated KC-130Rs are being bought to modernise this force. Three C-9Bs are used for high-speed transport duties. There are three Rockwell OV-10A squadrons for observation missions, a helicopter assault force of nine

Above: *US Army Grumman OV-1D with reconnaissance pod.* / Grumman

squadrons with 170 Boeing-Vertol CH-46F Sea Knights, and a further six squadrons with 100 CH-53D Sea Stallions. A batch of 33 updated CH-53E Super Stallions have been ordered for heavy-lift work. Six utility squadrons operate 116 Bell UH-1E/Ns while three attack squadrons have 74 Bell AH-1J SeaCobras, plus 57 TOW missile-armed Bell AH-1Ts on order with deliveries planned through 1979.

The USMC reserve force controlled by the 4th MAW has six A-4E/F Skyhawk squadrons, two F-4N squadrons, one OV-10 squadron, one KC-130 squadron, three CH-46D squadrons, three CH-53 squadrons, three UH-1E squadrons and one AH-1G squadron.

Title: United States Army
Headquarters: Washington
Strength: 10,000 aircraft

The world's largest helicopter operator, with more than 9,000 in service, the US Army is embarking on a major re-equipment programme which will continue well into the mid-1980s. To replace the large force of some 4,000 Bell UH-1s, the Service has selected the Sikorsky UH-60A Black Hawk as its future utility helicopter with initial procurement totalling 200 machines. Similarly, the Hughes AH-64A has been ordered to fulfil the Advanced Attack Helicopter requirement, replacing the Bell AH-1 HueyCobra, with procurement of the new type expected to reach 536 machines, including the three prototypes. As the AH-64 is not scheduled to reach Army units in any numbers until the mid-1980s, the service is updating large numbers of AH-1s from the existing G standard to the AH-1S Cobra, armed with TOW anti-tank missiles. Up to March 1979, 148 of the improved anti-armour Cobras have been ordered and by 1984, the US Army will have a fleet of 987 AH-1S helicopters in service, this figure including 290 interim AH-1Q versions.

The present US Army inventory includes 2,200 Bell OH-58A Kiowas for liaison purposes, 1,400 Hughes OH-6A Cayuse light observation machines and 54 Bell UH-1Hs recently delivered to supplement the UH-1 force, most of which have already been updated to H standard. Older types in use include Sikorsky H-34s and UH-19s, while medium and heavy-lift support for the ground forces is performed by 456 Boeing-Vertol CH-47A/B/C Chinooks (to be progressively updated to the latest CH-47D standard by 1993) and 80 CH-54 Tarhe skycranes. A large training force includes some 300 Hughes TH-55A primary helicopter trainers.

Fixed-wing aircraft operated by the US Army are limited to twin-engined types in size and their use extends from liaison and communications, to utility and training work. About 250 Grumman OV-1 Mohawk two-seat observation aircraft are in service, of 335 built in four variants, some operating as electronic EV-1s. Beech U-8 Seminole and U-21 Ute liaison aircraft are being supplemented by 80 Beech C-12A Super King Airs, named Hurons in Army service, and there are two DHC OV-18A Twin Otters flown by the Alaskan National Guard. Fixed-wing training is flown on 250 Cessna T-41s and 60 Beech T-42s.

Uruguay

Title: Fuerza Aerea Uruguaya
Headquarters: Montevideo
Strength: 80 aircraft

Few modern aircraft have been acquired by the FAU over the past few years due to a tight defence budget. The air force is organised into groups operating under the control of three Air Commands: Tactical Air Command with two Brigada Aerea, or Air Brigades, Training Air Command and Material Air Command. As Uruguay is a signatory of the 1947 Rio Pact, her air force has in the main been supplied

with aircraft from the United States, although some equipment has been bought from Brazil.

Air Brigade 1 controls Aviation Group 1 (Fighter) and operates six armed Lockheed AT-33As, the country's sole interceptor element following the US Government veto on the sale by Argentina of 20 North American F-86F Sabres to the FAU. Aviation Groups 3 and 4 (Transport), flying 12 Douglas C-47s, three Fairchild FH227s, two F27 Mk 100s and two Beech Queen Air liaison aircraft, also come under AB1 as does Aviation Group 6, which has four Embraer EMB-110 Bandeirante light transports for a dual civil/military role, plus one EMB-110B1 for photo/transport duties. Aviation Group 5 (Seek and Attack) is the other FAU combat unit and operates eight Cessna A-37Bs delivered in 1976.

Air Brigade 2 controls only one Group, No 1 (Tactical Reconnaissance), operating 10 North American T-6G Texans, 10 Beech AT-11s and six Cessna U-17 Skywagons. Helicopters include two Hiller UH-12s for SAR work, two Bell UH-1Hs and six UH-1Bs for troop carrying duties. Training Air Command has a Command and Air Staff School equipped with six Cessna T-41s and 25 T-34s received in 1977-78 from US stocks.

Aviacion Naval Uruguaya operates from two main shore bases at Punta del Este and Laguna del Sauce using three Grumman S-2A Trackers for ASW patrols. Supporting these are three Beech SNB-5s flown for training, a Beech T-34A for primary tuition, four T-6s, some Sikorsky SH-34s and a couple of Piper Super Cubs.

Above: *Cessna A-37B of the Uruguayan Air Force.*

Venezuela

Title: Fuerzas Aereas Venezolanas
Headquarters: Caracas
Strength: 260+ aircraft

Combat

To protect her mineral wealth, Venezuela has built up a substantial air arm equipped with modern aircraft operated on a high level of efficiency. The FAV has two combat groups, or Grupos, one encompassing the fighter squadrons, or Escuadrones, and the other the bomber squadrons. At the main fighter base of Barquisimento, Grupo 12 administers three fighter squadrons: Escuadrone 34, with 15 ex-Canadian AF Canadair CF-5As and four CF-5B trainers obtained in 1971; Escuadron 35, with 20 remaining North American/Fiat F-86K Sabres of 47 acquired from West Germany in the mid-1960s; and Escuadron 36, with nine Dassault Mirage IIIEV interceptors, four 5V ground-attack aircraft and two 5DV trainers delivered in 1973.

Grupo 13 operates the bomber element comprising two squadrons. Based at Maracay is Escuadron 39 operating 29 BAC Canberras (18 B2s, seven B(I)88s, two T84s and two PR83s) received between 1950 and 1960, the unit being divided into five flights, or Escuadrillas. The Canberras are currently (1978) undergoing a modernisation programme to extend their useful lives. The other unit is Escuadron 40 which is based at Barcelona and equipped with 16 Rockwell OV-10E Broncos delivered in 1973.

Transport and Support

Transport duties are performed by two units within the Grupo de Transporte based at Caracas. Escuadron 1 provides a heavy-lift element in the shape of five Lockheed C-130H Hercules delivered in 1971, and a secondary element flying 20 Douglas C-47s, some used as twin-engined trainers. Escuadron 2 has the survivors of 18 Fairchild C-123B Providers delivered in the mid-1950s and used for freight work. A VIP Flight has a single Boeing 737-200S received in 1976, and two HS748s, recently joined by a McDonnell Douglas DC-9

Grupo de Reconcimento flies helicopters from Caracas and Maracay comprising 15 Alouette IIIs, 10 Sikorsky UH-19s, 12 Bell UH-1D/Hs, six Bell 206B JetRangers and a Bell 206L LongRanger for VIP use. Fixed-wing liaison and communications duties are flown by 12 Cessna 182Ns and about nine Beech Queen Airs. It is reported that the Escuela de Aviacion Militar or School of Military Aviation, is to

Above: *Bell UH-1B of the Venezuelan Air Force.* / Bell

receive up to 20 Saab MFI-15/17 primary trainers and if substantiated, these will partially replace some 25 Beech T-34A Mentors used since 1959 when 41 were acquired by the FAV. A total of 24 Rockwell T-2D Buckeye advanced trainers are operated from Palo Negro, half the aircraft being fitted with weapons equipment for armament training and light strike duties. Twelve BAC Jet Provost T52s were delivered in 1962-63 and their present status remains obscure although they have been used for weapons training.

Servicio de Aviacion Naval is responsible for ASW duties along the country's coastline and for that purpose uses six ex-US Navy Grumman S-2E Trackers and four HU-16B Albatross amphibians. Three Douglas C-47s were replaced in 1978 by an ex-airline HS748 and a Beech King Air 90 which cater for transport duties, and there are two Bell 47Gs for liaison work. Two Agusta-Bell AB212ASW helicopters were ordered in 1977 to supplement the fixed-wing ASW machines.

Aviacion del Ejercito Venezolana is the army's air arm and is equipped with about 20 helicopters (10 Alouette IIIs, six Bell 47Gs and about two Sikorsky UH-19Ds) for liaison and AOP work.

Vietnam

Title: Vietnamese People's Air Force
Headquarters: Hanoi
Strength: 300 aircraft

The VPAF, originally the air arm of North Vietnam, now operates over what was South Vietnam, whose government collapsed in 1975. Border clashes with neighbouring Cambodia in 1978 are the result of ideological differences between the two communist regimes and have seen the limited use of air power by the Vietnamese. Large stocks of Western equipment and about 1,100 aircraft captured from the South Vietnamese remain in the country, although unofficially, Vietnam has indicated that it is willing to sell off some of the war booty.

The VPAF has a strike force of 10 Il-28 light bombers, two interceptor squadrons with 30 Chinese-built MiG-19s (Shenyang F-6s), six fighter-bomber units with about 80 MiG-17s, two strike units with 30 Su-7s, and four interceptor squadrons with 70 MiG-21s. Transports include 20 An-2s, four An-24s, 12 Il-14s, 20 Li-2s and a VIP Il-18. The helicopter element includes 12 Mi-4s, five Mi-6 heavy-lift machines plus some nine Mi-8s. A further 10 Mi-4s fly under Navy command for SAR duties.

Yemen

Title: Yemen Arab Republic Air Force
Headquarters: Sana'a
Strength: 35+ aircraft

After many years of limited Soviet economic and military aid, North Yemen appears to be moving towards closer ties with the West. In 1977, Saudi

Below: *Yemen Arab Air Force Short Skyvan 3M transport.* / Shorts

Arabia was given permission by the US Government to transfer four Northrop F-5B trainers to the country, and mooted at the time was the possibility of Saudi Arabia funding an order for 12 F-5E fighter-bombers for the YARAF. The current inventory comprises a strike/interceptor unit with 12 MiG-17s, and a light bomber squadron flying 16 Il-28s. The transport force is made up of two Short Skyvan 3Ms and a few Douglas C-47s and Il-14s, two AB205s and some Mi-4s.

Yemen, South

Title: Air Force of the South Yemen People's Republic
Headquarters: Aden
Strength: 50+ aircraft

A war with the Saudi-backed North Yemen erupted in February 1979 and Soviet equipment began arriving by air to add to that already held by this communist state. The air force has an interceptor squadron with 12 MiG-21s, a bomber unit with six Il-28s, and a ground-attack squadron with 15 MiG-17s. A transport unit has four Douglas C-47s, four Il-14s, and three An-24s. About eight Mi-8s and some Mi-4s are flying and three MiG-15UTIs are used for advanced training. A small detachment of Soviet Il-38 May patrol aircraft arrived in Aden in November 1978.

Yugoslavia

Title: Jugoslovensko Ratno Vazduhoplovstvo
Headquarters: Zemun
Strength: 688 aircraft

Combat

This progressive Communist country has an air arm flying equipment from both East and West, although its more modern combat aircraft have been supplied by the Soviet Union. Eight interceptor squadrons form two divisions flying about 110 MiG-21F/PF/Ms and fewer than 50 North American F-86D/K all-weather fighters. Two ground-attack divisions, with 12 squadrons, have 150 indigenous Soko Jastrebs and 30 Kraguj light attack aircraft. The Orao (Eagle) light strike/trainer, developed with Romania, is expected to join the Yugoslav Air Force's ground-attack element in the 1980s after a protracted development programme which began with the first flight of the prototype in 1974. Two reconnaissance squadrons have 15 Lockheed RT-33As and some Galeb/Jastrebs.

Transport and Support

For Presidential use there is a single Aérospatiale Caravelle 6N and two Boeing 727-200s, likely to be joined by a Rockwell 75A Sabreliner during 1978. Four Douglas DC-6Bs, 15 C-47s, 12 An-12s and 12 Il-14s form the fixed-wing transport force, together with an Il-18 and at least five Yak-40s for governmental duties. The YAF is receiving a number of An-26s from Russia to replace the fleet of C-47s. Helicopters total 10 Westland Whirlwinds, 18 Mi-4s, 20 Alouette IIIs, 12 Mi-8s, some Agusta-Bell AB205s and 132 licence-built Aerospatiale Gazelle anti-tank and liaison machines. For utility duties a substantial number of home-designed UTVA-66s are flown, and the training elements have 60 Galebs for basic tuition, 30 Lockheed T-33As and some MiG-15UTI two-seat advanced trainers.

The Yugoslav Navy operates Mi-8s in the transport/assault roles, some licence-built Gazelles and a few Kamov Ka-25 co-axial helicopters for ASW missions.

Below: *Used for training with the Yugoslav Air Force, the Soko Galeb.* / Rolls-Royce

Zaire

Title: Force Aérienne Zairoise
Headquarters: Kinshasa
Strength: 150 aircraft

Combat

The most sophisticated warplane in service with the FAZ is the Dassault Mirage fighter-bomber, for which orders were placed in 1973 totalling 14 single-seat 5Ms and three two-seat 5DM combat trainers. At least five are known to have been lost and reports have suggested that due to severe financial problems, no more than a handful of Mirages have actually been delivered up to mid-1978. However, those that are in use, serve with No 21 Wing, 2 Groupement Aérien Tactique, based at Kamina. Two other squadrons in this Group are assigned the ground-attack role, one equipped with about ten surviving Aermacchi MB326GBs of 17 delivered in 1971 and flying from Kamina and Kinshasa, and the other unit with about eight North American T-6G Texans; 20 Reims F337 Milirole counter-insurgency aircraft have been received and may have replaced the Texans. Eight single-seat MB326Ks were ordered in 1978 to supplement the two-seat versions.

Transport and Support

Tactical transport duties are performed by two DHC Caribous, three DHC Buffaloes and two Curtiss C-46s of the 22nd Wing from Kinshasa, while under the command of the 1st Groupement Aerien are two further transport squadrons. One of these units has seven Lockheed C-130H Hercules, delivered to the FAZ between 1971 and 1977, while the other operates ten Douglas C-47s, four C-54s and two DC-6s. All these units are based at Kinshasa but their operational state is unclear. A helicopter element has 15 Alouette IIIs, seven Bell 47Gs and seven SA330 Pumas (two more Pumas were lost in 1978 during a guerrilla attack on an airfield), plus a single Super Frelon for the personal use of the President. No 13 Wing at Kinshasa acts as the training element in the FAZ equipped with 23 Siai-Marchetti SF260MZs forming 131 Sqn, about 12 of 15 Cessna 150 Aerobats flown in the primary role, some of the MB326s acting in the advanced training role, and about a dozen of 15 Cessna 310Rs delivered for the dual twin training/liaison role.

Below: *Zaire Air Force Dassault Mirage 5DM two-seat trainer.* / AMD-BA

Zambia

Title: Zambian Air Force
Headquarters: Lusaka
Strength: 150 aircraft

Known as Northern Rhodesia before independence in 1964, Zambia was initially helped to form an air arm by the UK. More recently it has received both Italian and East European aid and the front-line elements reflect this. For strike/training duties, a squadron of 21 Aermacchi MB326GBs operate from Mbala, alongside four Yugoslav Jastreb light attack aircraft and two Galeb trainers. From Lusaka, eight Siai-Marchetti SF260Ms are used in the basic training role but have provision for underwing weapon pylons for the counter-insurgency task.

A transport unit operates seven DHC Buffalo STOL aircraft, 10 Douglas C-47s, five DHC Caribous, 10 Dornier Do28 Skyservants and seven DHC Beavers. With the transport unit at Lusaka is a VIP flight equipped with one HS748 and two Soviet-supplied Yak-40s. A helicopter fleet totals 25 Agusta-Bell AB205As, 17 Bell 47Gs, eight Alouette IIIs and one AB212; six Mil Mi-6 heavy-lift machines are reportedly in use. Utility duties and training are performed by 20 Saab MFI-17 Safaris delivered in 1976-77.

Addenda

Afghanistan: Since the socialist takeover in April 1978 the USSR is taking a more active role in the Afghan armed forces. The Czech L-39 Albatross trainers were delivered in 1978 and the transport force also includes about seven An-26s.

Albania has broken off relations with China.

Argentina has, since the shelving of the Mirage deal, purchased 26 Israeli-built versions known as Daggers and powered by Atar engines. The USA has halted delivery of the three Boeing Chinook helicopters.

Australia needs a total of 75 aircraft to replace their 84 Mirages. A decision is not expected before 1980 with an in-service date of 1984.

Austria: The expected announcement for the OKL's new combat aircraft has been deferred but information suggests the F-5E to be chosen with procurement expected c1980.

Bangladesh: Only about five MiG-21s remain airworthy.

Brazil: There are now 140 T-25 Universals for training and 20 have been ordered to make up for attrition. Embraer are to develop a new all-through turboprop trainer for the FAB designated T-27.

Bulgaria: Transports include An-14s; their helicopter regiments include Ka-26 liaison machines; utility duties are still performed by a number of An-2s.

Burma: The eight Douglas C-47 transports have been disposed of following receipt of the F27s. There are now nine Kaman HH-43B Huskies and 10 KV-107s in the helicopter fleet.

Cameroun now has two Dornier Do28s.

Chad Republic also has about 10 Alouette IIIs for liaison work.

Chile: The Ejercito de Chile received late in 1978 four Spanish Casa C212 Aviocar light transports, the type becoming the largest aircraft in Chilean Army service.

Columbia: The air force Transport Command has one (of three) C-130B Hercules.

Left: *An AV-8A Harrier takes off from USS Guam during interim sea control ship tests.* / Official US Navy

Cuba: The MiG-23s and MiG-27s are based at two airfields — San Julia in the west and Guinea near Havana. Cuba also possesses at least five An-26s.

Egypt: The 14 Lockheed C-130s are now being delivered. In 1978 four more Westland Commando 2Es were ordered.

Finland is actively seeking a replacement for their C-47s, of which they only have seven.

France: The Boeing KC-135F tankers are due to be re-engined with CFM56 turbofans. To extend their radar range a French AWACS is being developed based on the A300 Airbus and scheduled for service in 1982. The French Navy's carriers are due for retirement in 1990-92 and 42 of the updated 'Atlantic Nouvelle Generation' are to be ordered. Also reported in 1979 was a Naval requirement for four BA Coastguarder patrol aircraft. The French ALAT is going to receive 166 Gazelles, 110 modified to carry HOT. The first SA341F/HOT unit is the 3rd Regiment d'Heli based at Etain which formed in September 1978. A follow on order for the improved SA342M Gazelle has been placed, the number being 160, all armed with four HOT missiles.

Gabon: Only five Mirages have been ordered — three 5Gs and two two-seat 5DGs; delivery was expected in 1979.

Germany, West: The Alpha Jet 1A is entering service already and the first of 175 joined Waffenschule 50 in late 1978. The unit has, therefore, been redesignated JaboG49 since September 1978. The Marineflieger has since decided that Dornier should update their Breguet Atlantics with modern equipment over the next few years.

Great Britain: No 39 Squadron (Wyton) operates 17 Canberra PR9s. Since the withdrawal of British forces from Malta in 1978, No 13 Squadron with 14 Canberra PR7s has been based at Wyton. In 1979 No 216 Squadron reformed at Honington and moved to Lossiemouth equipped with Buccaneer S2Bs relinquished by No 809 Squadron RN, previously based on *Ark Royal*. Cottesmore accommodates a tri-national training unit for Tornados and a second such unit will be formed at Honington in 1980 and be called the Anglo-German Weapons Unit. No 38 Group, to make up for attrition, is purchasing seven more Pumas. 30 of the Hercules C1 Wing aircraft will be 'stretched'. The RAF Support Command Gnat T1s have been replaced by Hawks. The multi-engine training unit based at Leeming is also due to move to Finningly. The Fleet Air Arm's AS cruisers are HMS *Invincible*, HMS *Illustrious* (launched in December 1978) and HMS *Ark Royal*. HMS *Bulwark* has been restored to full operational status. No 750 Squadron's eight Sea Prince T1s have been replaced by Jetstream T2s in the observer training role. The AAC's first Lynx unit is No 654 Squadron based at Minden. No 660 Squadron's 11 Flight now flies Scouts.

Greece has now received their tactical-recce unit's F-4Es.

Honduras: Of the 10 T-28 Trojans, eight are ex-Moroccan examples received in August 1978.

Israel has indirectly supplied 11 of its 45 Bell 205s to Rhodesia leaving their own total 34. The SNECMA Atar 9C-powered 'Neshers' are believed to be the aircraft sold to Argentina in 1978.

Japan: As yet no order for No 111 Squadron's Sikorsky RH-53D requirement has been placed.

Lebanon also has an Aero Commander for transport work.

Libya: A two-seat MiG-25U of the Russian reconnaissance unit was lost in the sea in November 1978.

Madagascar has recently formed a combat element with the aid of North Korea who have supplied the air arm with eight MiG-17s.

Malaysia: No 12 'Tiger' Squadron has lost two F-5Es in accidents.

Mexico: The Light Transport Group also has some Piper Aztecs.

Mozambique received three Mi-8s from the USSR in 1978.

Netherlands have decided on the Lockheed P-3C Orion to replace their Neptunes and an order for 13 was placed in December 1978 with delivery planned for 1981.

New Zealand plans to replace its No 42 Squadron Devons with three ex-Air New Zealand F27s.

Nigeria operates -21MF versions of the MiG-21. To replace their L-29s the Nigerian Air Force has ordered 12 Alpha Jets from Dornier with delivery due in 1981-82.

Norway has taken up the option on the two extra Westland Lynxes. Recently the RNoAF has purchased five ex-Swedish Safirs to take their training inventory to more than 20 aircraft.

Panama purchased a single Short Skyvan 3M in 1978.

Paraguay: In 1978 six Cessna A-37Bs were supplied to the FAP from USAF sources.

Philippines: As expected 25 LTV F-8H Crusaders have replaced the Sabres of the 7th TFS while the 9th TFS has reverted to a training unit equipped with its Mentors and six remaining F-86Fs. 10 LTV F-8Hs are being stored for use as spares.

Portugal: Front-line air defence is performed by Escuadra 201. Advanced training work is flown by Escuadra 103.

Rhodesia: No 7 Squadron's Alouette IIIs have been reported as being as many as 66 and in 1978 it was revealed that 11 Bell 205s were acquired for No 7 Squadron from Israel. They are being used as gunships and utility machines. Also acquired via an order for the Comores Islands were 20 SF260W Warriors.

Saudi Arabia has 40 Indonesia-built Casa C212 Aviocars on order for light transport duties.

Singapore no longer operates the Alouette IIIs and has 17 Bell UH-1Hs.

Soviet Union: The aircraft carrier *Minsk* entered service in 1979.

Spain: The 48 Mirages on order are made up of 42 F1CE interceptors and six F1BE trainers; three more Orions are expected to be ordered in 1979. Nine SA330T Pumas are on order for SAR duties with delivery due in 1979.

Sudan has ordered 10 F-5Es and two F-5Fs.

Taiwan: America's official recognition of mainland China has forced a break in diplomatic relations but outstanding arms agreements between the two countries will be honoured. Two Beech Super King Airs have been bought by Taiwan for navaid calibration.

Tanzania has recently acquired a Fokker F28 Mk 3000 for government use.

Thailand has lost one of its Royal Flight's existing Swearingen Merlin IVs, and three more are on order in 1978 for photographic work.

Turkey has, reportedly, three Viscounts for use at Etimesgut.

USA: Of the three A-10A wings based in Europe, one is based in England and the other two in Germany are at Sembach and Leipheim. The 34 Beech C-12s for air attache use have been delivered. The USMC Harrier development work has been temporarily halted by the 1979 budget.

Uruguay: Air Brigade 1 also flies one Cessna 310L; Air Brigade 2 no longer has its 10 Beech AT-11s.

Vietnam: The Chinese invasion of Vietnam in February 1979 which followed Vietnam's invasion of Cambodia in the January found the VPAF using a mixture of eastern and western equipment including one squadron that operated a mixture of F-5Es and MiG-21s.

Yemen: A war with neighbouring South Yemen has put some urgency on the acquisition of more up-to-date arms and as well as the ordered F-5Es a batch of C-130 Hercules is likely to be supplied by the USA.

Zambia: At least two of the Aermacchi MB326GBs have been lost.

Index

A successful French export with over 500 built or on order was the Mirage F1. These are French Air Force F1Cs. / AMD-BA